Fluid Mechanics and Hydraulic Machinery

Fluid Mechanics and Hydraulic Machinery

Branden Harrison

CLANRYE
INTERNATIONAL
www.clanryeinternational.com

Clanrye International,
750 Third Avenue, 9th Floor,
New York, NY 10017, USA

ISBN: 978-1-64726-647-9

Cataloging-in-Publication Data

Fluid mechanics and hydraulic machinery / Branden Harrison.
p. cm.
Includes bibliographical references and index.
ISBN 978-1-64726-647-9
1. Fluid mechanics. 2. Hydraulic machinery. 3. Hydraulics. 4. Hydraulic engineering--Instruments.
I. Harrison, Branden.

TA357 .F58 2023
620.106--dc23

For information on all Clanrye International publications
visit our website at www.clanryeinternational.com

CLANRYE INTERNATIONAL

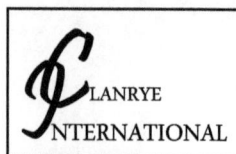

Contents

Preface

Fluid mechanics refers to the branch of physics that studies the mechanics of forces acting on fluids such as plasmas, gases and liquids. It is used in many disciplines such as geophysics, meteorology, chemical and biological engineering, mechanical engineering, oceanography, biology, civil engineering and astrophysics. It is classified into two parts including fluid dynamics, which studies the effect of forces on fluid motion, and fluid statics, which studies fluids at rest. Hydraulic machines work by utilizing liquid fluid power to perform their work, such as heavy construction vehicles. These machines generally pump hydraulic fluid to numerous hydraulic cylinders and hydraulic motors throughout the machine and it gets pressurized based on the resistance. From theories to research to practical applications, studies related to all contemporary topics of relevance to fluid mechanics and hydraulic machinery have been included in this book. It will provide comprehensive knowledge to the readers.

All of the data presented henceforth, was collaborated in the wake of recent advancements in the field. The aim of this book is to present the diversified developments from across the globe in a comprehensible manner. The opinions expressed in each chapter belong solely to the contributing authors. Their interpretations of the topics are the integral part of this book, which I have carefully compiled for a better understanding of the readers.

At the end, I would like to thank all those who dedicated their time and efforts for the successful completion of this book. I also wish to convey my gratitude towards my friends and family who supported me at every step.

Branden Harrison

1

Fluid Statics

1.1 Fluid Statics: Dimensions and Units

A dimension is a measure of a physical variable (without numerical values), while a unit is a way to assign a number or measurement to that dimension. For example, length is a dimension but it is measured in units of feet (ft.) or meters (m).

There are three primary unit systems in use are:

- The International System of Units (SI units, from 'Le System International d'Unites', more commonly called as metric units).
- The English Engineering System of Units (commonly called English units).
- The British Gravitational System of Units (BG).

Table: Some of the dimensions and units of various physical quantities.

Terminology	Dimensions	Imperial Units (USCS)	SI-units
Acceleration due to Gravity	L/T^2	ft/s^2	m/s^2
Area	L^2	ft^2	m^2
Chezy Roughness Coefficient	$L^{1/2}/T$	$ft^{1/2}/s$	$m^{1/2}/s$
Critical Depth	L	ft	m
Density	FT^2/L^4	$lb\,s^2/ft^4$	$N\,s^2/m^4$
Depth	L	ft	m
Depth in Open Channel	L	ft	m
Diameter	L	ft	m
Distance from Solid Boundary	L	ft	m
Flow Rate	L^3/T	ft^3/s	m^3/s

Force	F	lb	N
Force due to Pressure	F	lb	N
Hazen Williams Roughness Coefficient	$L^{0.37}/T$	$ft^{0.37}/s$	$m^{0.37}/s$
Head Loss Due to Friction	L	ft	m
Head of Height	L	ft	m
Head of Weir	L	ft	m
Height above Datum	L	ft	m
Hydraulic Radius	L	ft	m
Kinematic Viscosity	L^2/T	ft^2/s	m^2/s
Length	L	ft	m

Table: Let us compare solids and liquids.

S. No.	Solids	Liquids
1.	It has its own shape and undergoes an infinitesimal change in volume under pure compressive load.	It does not have their shape.
2.	It offer's resistance to change in shape without change in volume under the application of tangential force.	They do not offer any resistance to change in shape when a deforming tangential force is applied.
3.	Inter-molecular cohesive forces are large.	Inter molecular cohesive forces are Smaller.
4.	Examples: Bricks, Steel, Word, etc.	Examples: Water, Milk, Kerosene and etc.

1.1.1 Physical Properties of Fluids

Mass Density

Mass Density is defined as the ratio of mass per unit volume. It is represented by ρ. Its unit is kg/m3. Mathematically it is written as:

$$\rho = m/V$$

Specific Weight

Specific weight is the weight possessed by unit volume of a fluid. It is denoted by 'w'. Its unit is N/m3. Specific weight varies from place to place due to the change of acceleration due to gravity (g).

Specific weight, w = Weight/Volume N/m^3

Table: Comparison of Mass Density and Specific Weight.

S. No.	Mass Density	Specific Weight
1.	It is defined as the mass per unit volume.	It is defined as the weight per unit volume at the standard conservative and pressure.
2.	It is devoted by 'ρ' (rho)	It is denoted by (w).
3.	It is also called as Density or Specific mass.	It is also called as weight density.
4.	$\rho = \frac{M}{V}$	$W = \frac{w}{V} = \frac{Mg}{V} = \rho g$ $W = \rho g$

Specific Volume

Specific volume is the volume of a fluid (V) occupied per unit mass (m). It is the reciprocal of density. Specific volume is denoted by the symbol 'v'.

Specific Volume, v=Volume/mass=1/m^3/kg

Specific Gravity

Specific gravity is the ratio of specific weight of the given fluid to the specific weight of standard fluid. It is denoted by the letter 'S'. It has no unit.

Specific Gravity, S = Specific Weight of Given Fluid/Specific Weight of Standard Fluid

Specific gravity may also be defined as the ratio between density of the given fluid to the density of standard fluid.

Problems

1. The specific gravity of oil is 0.8 which produces a pressure of 120 KN/m2. Let us determine the corresponding depth of water.

Solution:

Given data:

- Specific Gravity of oil = 0.8
- Pressure Intensity (P) = 120 kN/m^2

To find: Corresponding Depth of Water.

Formula to be used:

- Specific gravity of oil × Specific weight of water
- $V = \frac{\text{Viscosity}}{\text{Density}} = \frac{\mu}{\rho}$

Specific weight of the oil (w):

W = Specific gravity of oil × Specific weight of water

$= 0.8 \times 9.81$

$\Rightarrow 7.848\ kN/m^3$

Intensity of pressure $(P) = \omega\, h$

$120 = 7.848 \times h$

$h = 15.29$ m of water

Kinematic Viscosity

Kinematic Viscosity is defined as the ratio between the dynamic viscosity and density of fluid. It is denoted by 'V'. Mathematically,

$$V = \frac{\text{Viscosity}}{\text{Density}} = \frac{\mu}{\rho}$$

Viscosity

Viscosity is defined as the property of a fluid which offers resistance to the movement of one layer of fluid over another adjacent layer of the fluid.

$$\text{unit} \rightarrow \frac{NS}{m^2} = 10\,\text{Poise}$$

Effect of Temperature on Viscosity

The temperature on Viscosity is affected by temperature. The viscosity of liquids decreases but that of gases increases with increases in temperatures. This is due to the reason that in liquids the shear stress is due to the intermolecular cohesion which decreases with increases of temperature.

In gases the intermolecular cohesion is negligible and the shear stress is due to exchange of moments of the molecules, normal to the direction of motion. The molecular activity increases with rise in temperature and so does the viscosity of gas.

For liquids:

$$\mu_T = Ae^{B/T}$$

For gases:

$$\mu_T = \frac{bT^{1/2}}{1+a/T}$$

Where,

- μ_T = Dynamic viscosity at absolute temperature T.
- A, B = Constants (for a given liquid).
- a, b = Constants (for a given gas).

Effect of Pressure on Viscosity

The viscosity under ordinary conditions is not appreciably affected by the changes in pressure. However, the viscosity of some oils has been found to increase, with increase in pressure.

Newton's Law of Viscosity

Newton's law of viscosity states that "the shear stress (τ) on a fluid element layer is directly proportional to the rate of shear strain".

$$\text{i.e.,} \tau \alpha \frac{du}{dy}$$

The constant of proportionality is called the co-efficient of viscosity. Mathematically it is given by equation:

$$\tau = \mu \frac{du}{dy}$$

Where,

- $\tau \rightarrow$ Shear stress
- $\mu \rightarrow$ Co-efficient of viscosity
- $\frac{du}{dy} \rightarrow$ rate of shear strain (velocity gradient)

The fluids which obey the above relation are known as newtonian fluids and the fluids which do not obey the above relation are called Non-Newtonian fluids.

When two layers of a fluid, a distance 'dy' apart, mover one over the other at different velocities, say 'u' and 'u + du' as shown in figure. The viscosity together with relative velocity causes a shear stress acting between the fluid layers.

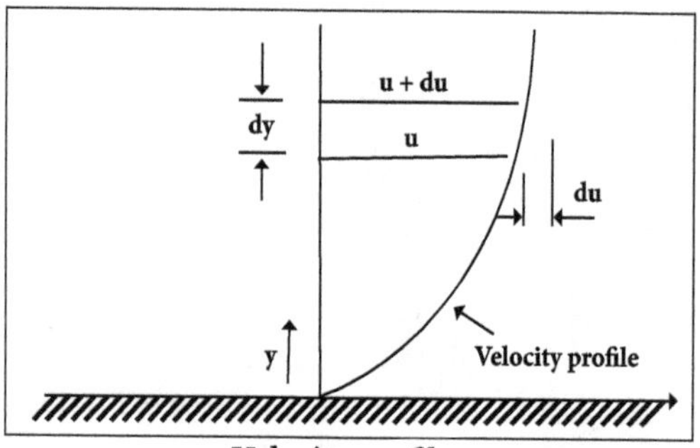

Velocity profile.

2. Let us determine the depth of oil, specific gravity 0.8, will produce a pressure of 120 KN/m^2? What would be corresponding depth of water?

Solution:

Given data:

- Specific Gravity of oil = 0.8
- Pressure Intensity (P) = 120 kN/m2

To find: The depth of oil.

Formula to be used:

- Specific gravity of oil × Specific weight of water
- Intensity of pressure (P) = ω h

Specific weight of the oil (w):

- w = Specific gravity of oil × Specific weight of water
- $= 0.8 \times 9.81$
- $\Rightarrow 7.848\ kN/m^3$

Intensity of pressure (P) = ω h

$$120 = 7.848 \times h$$

$$h = 15.29 \text{ m of water}$$

Surface Tension and Capillarity

The surface tension is defined as the tensile force acting on the surface of a liquid in contact with a gas or on the surface between two immiscible liquids such that the contact surface behaves like a membrane under tension.

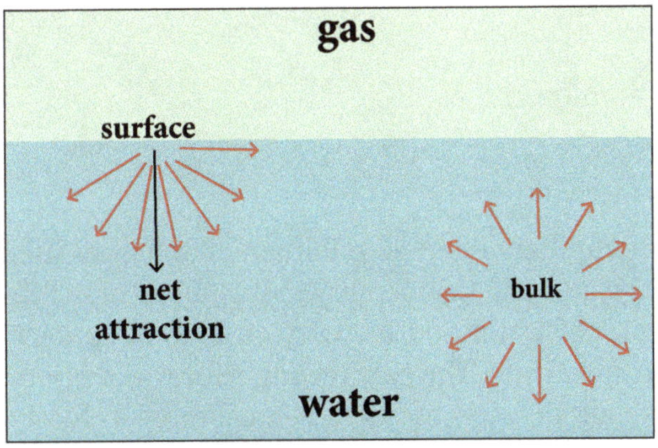

Surface Tension.

The capillarity is a phenomenon by which a liquid [depending upon its specific weight] rises into a thin glass tube above or below its general level. This is due to the combined effect of cohesion and adhesion of liquid particles.

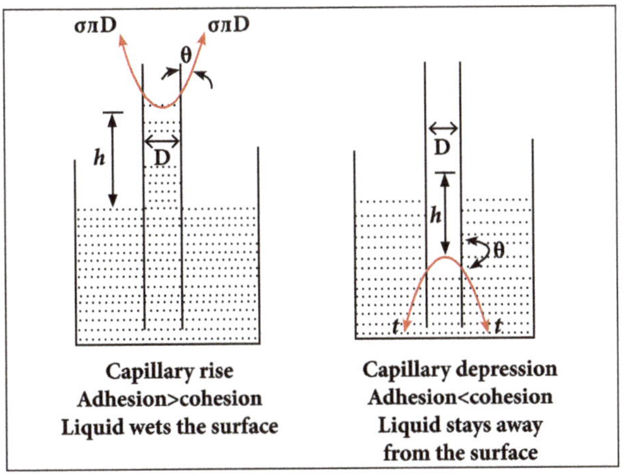

Capillarity.

3. The soap bubble is formed when the pressure inside is 5 N/m above the atmospheric pressure. If surface tension in the soap bubble is 0.0125 N/m, et us find the diameter of the bubble formed.

Solution:

Given data:

- $P = 5\ N/m^2$
- Surface tension, $\sigma = 0.0125 N/m$

To find: Diameter of the bubble, d.

Formula to be used:

$$P = \frac{8\sigma}{d}; d = \frac{8\sigma}{P}$$

$$d = \frac{8 \times 0.0125}{5} = 0.02\,m$$

$$d = 2\,cm$$

4. Let us calculate the capillary effect in millimeters in a glass tube of 4 mm diameter, when immersed in (1) water and (2) mercury the temperature of the liquid is 20°C. The value of surface tension of water and mercury at 20°C in contact with air are 0.0735 N/m and 0.51 N/m respectively. The contact angle for water θ = 0 and for mercury θ = 130°. Take specific weight of water at 20°C as equal to 9790 N/m3 and specific gravity of mercury is 13.6.

Solution:

Given:

- Diameter of the glass tube (d) = 4 mm= 0.004 m.
- Surface tension at 20°C.
- σ water = 0.0735 N/m.
- σ mercury = 0.05 N/m.
- Specific weight of water = 9790 N/m^3.

To find:

- Capillary effect of water.
- Capillary effect of mercury.

Formula to be used:

$$h = \frac{4\sigma\cos\theta}{\omega d}$$

The rise or depression 'h' of a liquid in a capillary tube is $h = \frac{4\sigma\cos\theta}{\omega d}$.

1. Capillary effect of water:

$$h = \frac{4 \times 0.07351 \cos 0^\circ}{9790 \times 0.004} \quad \theta_{water} = 0$$

$$= 0.751 \times 10^{-3m}$$

$$h = 7.51\,mm \; [Rise]$$

2. Capillary effect of mercury:

$$h = \frac{4 \times 0.051 \times \cos 130^\circ}{(13.6 \times 97ao) \times 0.004} \quad \theta_{mercury} = 130^\circ$$

$$= -2.46 \times 10^{-3m}$$

$$= -2.46\,mm$$

$$h = 2.46\,mm \; [Depression]$$

Vapor Pressure

The vapor pressure or equilibrium vapor pressure is defined as the pressure exerted by a vapor in thermodynamic equilibrium with its condensed phases (solid or liquid) at a given temperature in a closed system.

The equilibrium vapor pressure is an indication of a liquid's evaporation rate. It related to the tendency of particles to escape from the liquid (or a solid). A substance with a high vapor pressure at normal temperatures is often referred to as volatile.

The vapor pressure of any substance increases non-linearly with temperature according to the Clausius Clapeyron relation. The atmospheric pressure boiling point of a liquid (also known as the normal boiling point) is the temperature at which the vapor pressures same the ambient atmospheric pressure.

With any incremental increase in that temperature, the vapor pressure becomes sufficient to overcome atmospheric pressure and lift the liquid to form vapor bubbles inside the bulk of the substance. Bubble formation deeper in the liquid requires a maximum pressure and therefore higher temperature, because the fluid pressure increases above the atmospheric pressure as the depth increases.

The vapor pressure that a single component in a mixture contributes to the total pressure in the system is called partial pressure. For the example, air at sea level and saturated with water vapor at 20 °C, has partial pressures of about 2.3 Kpa of water, 78 kpa of nitrogen, 21 Kpa of oxygen and 0.9 Kpa of argon.

1.2 Atmospheric Gauge and Vacuum Pressure

1. Absolute pressure (P_{abs}) - The actual pressure at a given position is called the absolute pressure. It is measured relative to absolute vacuum (i.e., absolute zero pressure).

$$P_{abs} = P_{atm} + P_{gauge}$$

2. Gauge pressure (P_{gauge}) - Gauge pressure is the pressure relative to the atmospheric pressure. It can also be expressed as how much of above or below is the pressure with respect to the atmospheric pressure.

$$P_{gauge} = P_{abs} - P_{atm}$$

3. Vacuum pressure (P_{vac}) - The pressures under atmospheric pressure are called vacuum pressures and are measured by vacuum gages that indicate the difference between the atmospheric pressure and the absolute pressure.

$$P_{vac} = P_{atm} - Pa_{bs}$$

4. Atmospheric pressure (P_{atm}) - The atmospheric pressure is the pressure that an area experiences due to the force exerted by the atmosphere. For calculations, the pressure used is the pressure at sea level. Typically, the quantity used for engineering calculations is 1 atm or101 kPa.

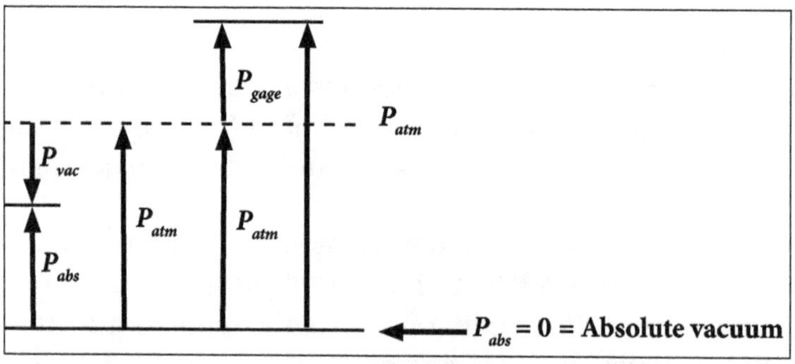

The actual pressure at a given position is called the absolute pressure and it is measured relative to absolute vacuum (i.e., absolute zero pressure). Most of pressure measuring devices are calibrated to read zero in the atmosphere and so they indicate the difference between the absolute pressure and the local atmospheric pressure. This difference is called the gauge pressure. Pressures below atmospheric pressure are called vacuum pressures and are measured by vacuum gauges that indicate the difference between the atmospheric pressure and the absolute pressure. The other pressure gauges, the gauge used to measure the air pressure in an automobile tire reads the gauge pressure. Therefore, the common reading of 32 psi (2.25 kgf/cm²) indicates a pressure of 32 psi above the atmospheric pressure.

1.3 Measurement of Pressure

Manometer

A manometer is a device which is used to measure the pressure of a fluid by balancing it against a column of liquid. The different types of manometers are given:

U-Tube Manometer

It consist a U – shaped bend whose one end is attached to the gauge point 'A' and the other end is open to the atmosphere. It is capable of measuring both positive and negative (suction) pressures. It contains liquid of specific gravity greater than that of a liquid of which the pressure is to be measured.

Pressure at A is $P = \gamma_2 h_2 - \gamma_1 h_1$

Where, 'γ' is Specific weight, 'P' is Pressure at A.

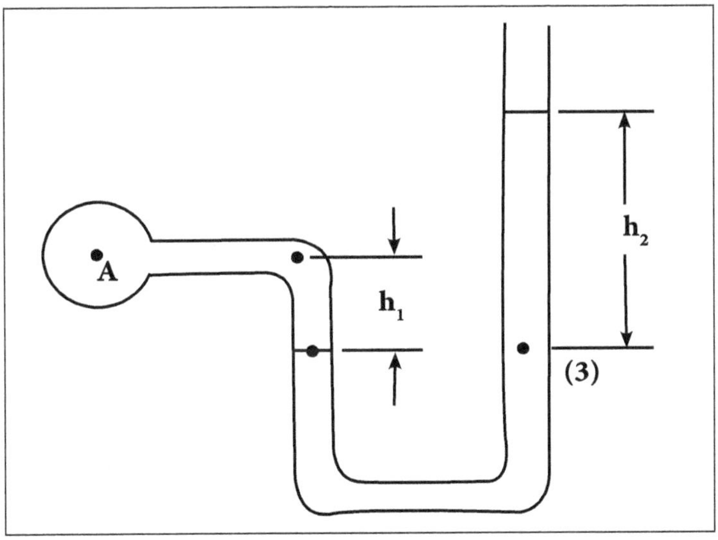

U-Tube Manometer.

Differential U-Tube Manometer

A U-Tube manometric liquid heavier than the liquid for which the pressure difference is to be measured and is not immiscible with it is measured using Differential manometer.

Pressure difference between A and B is given by equation:

$$P_A - P_B = \gamma_2 h_2 + \gamma_3 h_3 - \gamma_1 h_1$$

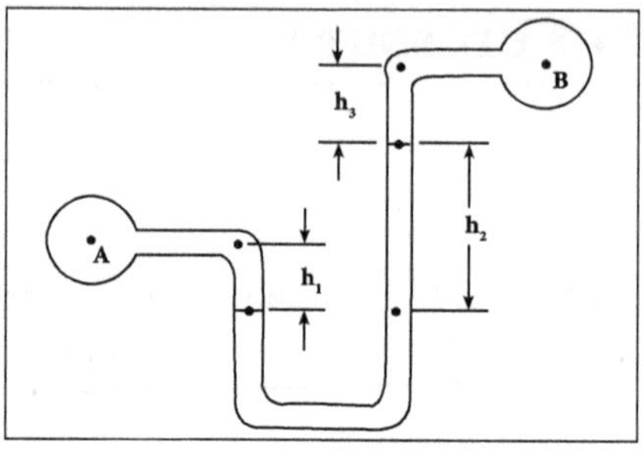

Differential U-Tube Manometer.

Inverted U-Tube Manometer

Inverted U-Tube manometer consists of an inverted U – Tube containing a light liquid. It is used to measure the differences of low pressures between two points where better accuracy is required. It consists of an air cock at top of manometric fluid type.

Pressure difference can be calculated from equation $P_1 - \rho_1 \times g \times H_1 - \rho m \times g(H_2 - H_1)$ $= P_2 - \rho_{2x} \times gH_2$

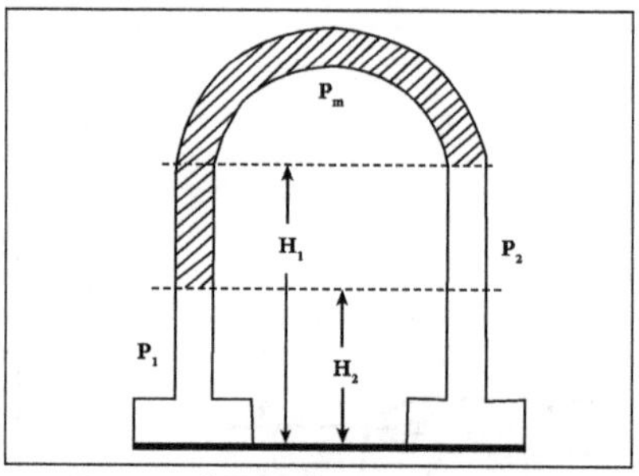

Inverted U-Tube Manometer.

Piezometer

Piezometer is the simplest forms of manometers. It is used for measuring moderate pressures of liquids. The piezometer consists of a glass tube, inserted in the wall of a vessel or of a pipe. The tube extends vertically upward to such a height that liquid can freely rise in it without overflowing. The pressure at any point in the liquid is indicated by the height of the liquid in the tube above that point.

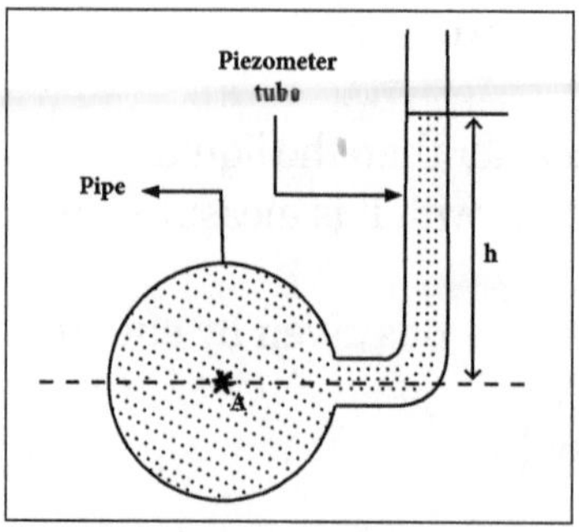

Pressure at point A can be computed by measuring the height to which the liquid rises in the glass tube.

The pressure at point A is given by:

$$p=wh$$

Where, W - is the specific weight of the liquid.

Limitations of Piezometer

- Piezometers can measure gauge pressures only. It is not suitable for measuring negative pressures.
- Piezometers cannot be employed when large pressures in the lighter liquids are to be measured since this would require very long tubes, which cannot be handled conveniently.
- Gas pressures cannot be measured with piezometers, because a gas forms no free surface.

Advantages of Manometers

- This is very simple.
- No calibration is required - The pressure can be calculated from first principles.

Disadvantages of Manometers

- Slow response - only really useful for very slowly varying pressures - no use at all for fluctuating pressures.
- For the "U" tube manometer two measurements must be taken simultaneously to get the h value. This may be avoided by using a tube with a much larger cross-sectional area on one side of the manometer than the other.
- It is often difficult to measure small variations in pressure - a different manometric fluid may be required - alternatively a sloping manometer may be employed, it cannot be used for very large pressures unless several manometers are connected in series.
- For very accurate work the temperature and relationship between temperature and r must be known.

Problems

1. The pressure in a tank is measured with a manometer by measuring the differential height of the manometer fluid. Let us determine the absolute pressure in the tank for two cases: the manometer arm with the (a) higher and (b) lower fluid level being attached to

the tank. Given that the fluid in the manometer is incompressible, the specific gravity of the fluid is given to be SG = 1.25 and the density of water at 32 °F is 62.4 lbm/ft^3.

Solution:

Given:

- SG = 1.25
- The density of water at 32°F is 62.4 lbm/ft^3

To find:

- The manometer arm with the Higher.
- Lower fluid level being attached to the tank.

Formula to be used:

- $P_{abs} = P_{atm} + P_{gauge}$

The density of the fluid is obtained by multiplying its specific gravity by the density of water:

$$\rho = SG \times \rho_{H_2O} = (1.25)(62.4\,lbm/ft^3) = 78.0\ lbm/ft^3$$

$$P_{gauge} = P_{abs} - P_{atm}$$

The pressure difference corresponding to a differential height of 28 in between the two arms of the manometer is:

$$\Delta P = \rho gh = (78\ lbm/ft^3)(32.174\ ft/s^2)(28/12\ ft)$$

$$\left(\frac{1\,lbf}{32.174\,lbm \cdot ft/s^2}\right)\left(\frac{1\,ft^2}{144\,in^2}\right) = 1.26\ psia$$

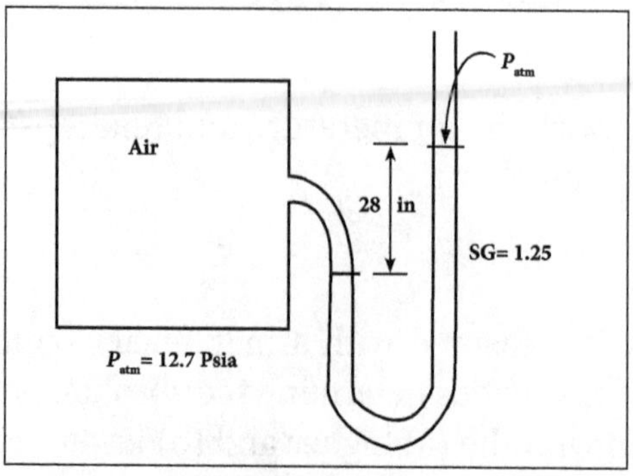

Then the absolute pressures in the tank for the two cases become:

- The fluid level in the arm attached to the tank is higher (vacuum):

$$P_{abs} = P_{atm} - P_{vac} = 12.7 - 1.26 = 11.44 \text{ psia} \cong 11.4 \text{ psia}$$

The fluid level in the arm attached to the tank is lower:

$$P_{abs} = P_{gauge} + P_{atm} = 12.7 + 1.26 = 13.96 \text{ psia} \cong 14.0 \text{ psia}$$

2. The pressure in a pressurized water tank is measured by a multi-fluid manometer. The densities of mercury, water and oil are given to be 13,600, 1000 and 850 kg/m3 respectively. Let us determine the gauge pressure of air in the tank.

Solution:

To find: The gauge pressure of air in the tank.

The air pressure in the tank is uniform (i.e., its variation with elevation is negligible due to its low density) and thus we can determine the pressure at the air water interface.

Starting with the pressure at point 1 at the air-water interface and moving along the tube by adding (as we go down) or subtracting (as we go up) the ρgh terms until we reach point 2 and setting the result equal to P_{atm} since the tube is open to the atmosphere gives:

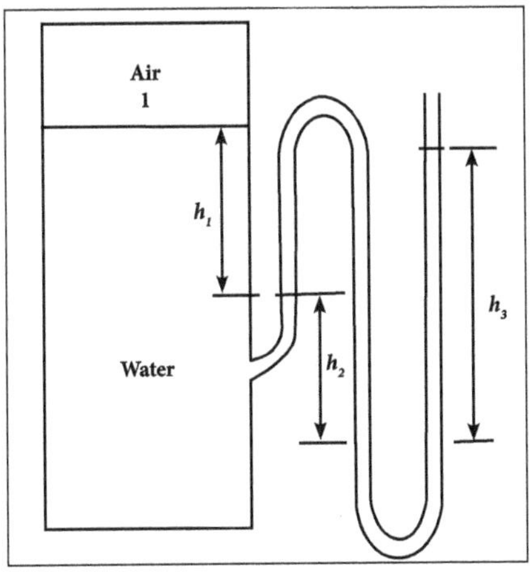

$$P_1 + \rho_{water} g h_1 + \rho_{oil} g h_2 - \rho_{mercury} g h_3 = P_{atm}$$

Solving for P_1

$$P_1 = P_{atm} - \rho_{water} g h_1 - \rho_{oil} g h_2 + \rho_{mercury} g h_3$$

or,

$$P_1 - P_{atm} = g\left(\rho_{mercury} h_3 - \rho_{water} h_1 - \rho_{oil} h_2\right)$$

Noting that $P_{1,gage} = P_1 - P_{atm}$ and substituting,

$$P_{1,gage} = (9.81\ m/s^2)\left[(13{,}600\ kg/m^3)(0.46\,m) - (1000\ kg/m^3)(0.2\,m)\right.$$

$$\left. -(850\ kg/m^3)(0.3\,m)\right]$$

$$\left(\frac{1\,N}{1\,kg \cdot m/s^2}\right)\left(\frac{1\,kPa}{1000\,N/m^2}\right)$$

$$= 56.9\ kPa$$

1.4 Pascal's Law and Hydrostatic Law

Pascal's Law

Pascal's law states that "The Intensity of pressure at any point in a liquid at rest is the same in all diversions".

Proof

Let us consider a small wedge shaped element ABC of a liquid as shown in the figure:

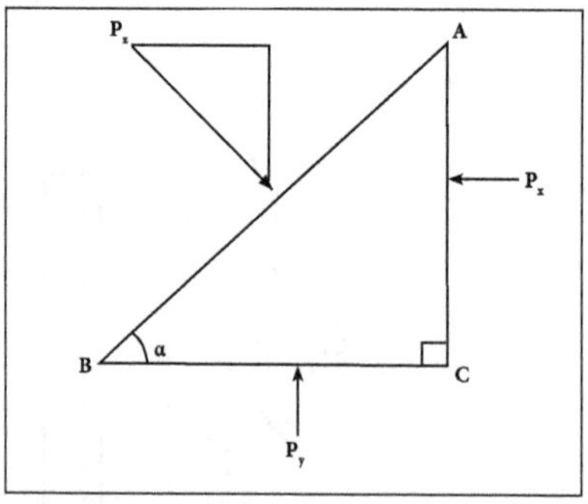

Wedge shaped element ABC of a liquid.

Let,

- p_x = Intensity of horizontal pressure on the element of liquid.
- p_y = Intensity of vertical pressure on the element of liquid.
- p_z = Intensity of pressure on the diagonal of the right angled triangular element.
- α = Angle of the element of the liquid.
- P_X = Total Pressure on the vertical side, AC of the liquid.
- P_Y = Total Pressure on the horizontal side, BC of the liquid.

- P_Z = Total Pressure on the diagonal, AB of the liquid.

Now,

$$P_X = p_x \times AC \quad ...(1)$$

$$P_Y = p_y \times BC \quad ...(2)$$

$$P_Z = p_z \times AB \quad ...(3)$$

As the element is at rest, therefore the sum of horizontal and vertical component of the liquid pressure must be equal to zero.

Resolving forces horizontally:

$$P_z \sin \alpha = P_x$$

$$P \times AB \times \sin \alpha = P_x \times AC \quad [\because P_Z = p_z \times AB]$$

But,

$$AB \sin \alpha = AC$$

$$\therefore P_Z = P_x \quad ...(4)$$

Resolving forces vertically:

$$P_z \cos \alpha = p_y - W$$

Where, W = Weight of the liquid element.

Since the element is very small, neglecting its weight and we get:

$$p_z \cos \alpha = p_y$$

Or,

$$p_z \cdot AB \cos \alpha = p_y \times BC$$

But,

$$AB \cos \alpha = BC$$

$$\therefore P_z = P_y \quad ...(5)$$

From equation 4 and 5 we get:

$$P_x = P_y = P_z$$

Which is independent of 'α'.

1.5 Buoyancy and Floatation: Meta Center and Stability of Floating Body

Buoyancy

Buoyant force is the hydrostatic lift due to the net vertical component of the hydrostatic pressure experienced by the body when it is wholly or partially immersed in a fluid and this phenomenon is called "Buoyancy".

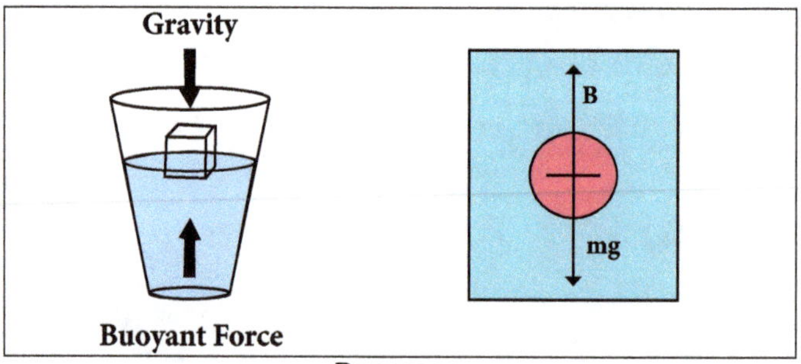

Buoyancy.

The upward force exerted by the fluid on the body when the body is immersed in a fluid or floating on a fluid is called buoyancy. This upward force is equal to the weight of the fluid displaced by the body.

Center of Buoyancy

A point through which the force of buoyancy is supposed to act is called Center of buoyancy. As the force of buoyancy is a vertical force and is equal to the weight of the fluid displaced by the body, the Center of Buoyancy will be the center of the fluid displaced.

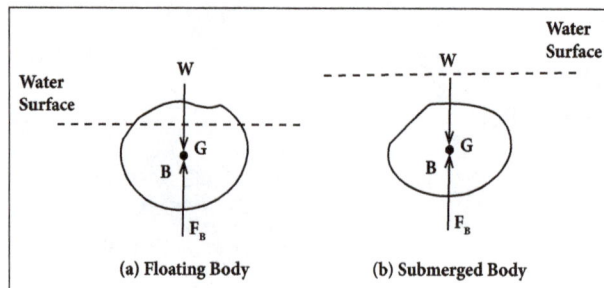

(a) Floating Body and (b) Submerged Body.

Archimedes Principle

The Buoyant Force (F_B) is equal to the weight of the liquid displaced by the submerged body and acts vertically upwards through the centroid of the displaced volume.

Net weight of the submerged body = Actual weight – Buoyant force

The buoyant force on a partially immersed body is also equal to the weight of the displaced liquid. It depends upon the density of the fluid and submerged volume of the body. For a floating body in static equilibrium, the buoyant force is equal to the weight of the body.

Stability of Floating Body

- The floating body is STABLE if, when it is displaced, it returns to equilibrium.
- The floating body is UNSTABLE if, when it is displaced, it moves to a new equilibrium.

It considers a floating body tilted by an angle $\Delta\theta$, as shown below. For the untilted body the point G is the centre of gravity of the body where the body weight, W, acts. The point B is the centre of buoyancy (the centroid of the displaced volume of fluid) where the upward buoyancy force, FB, acts.

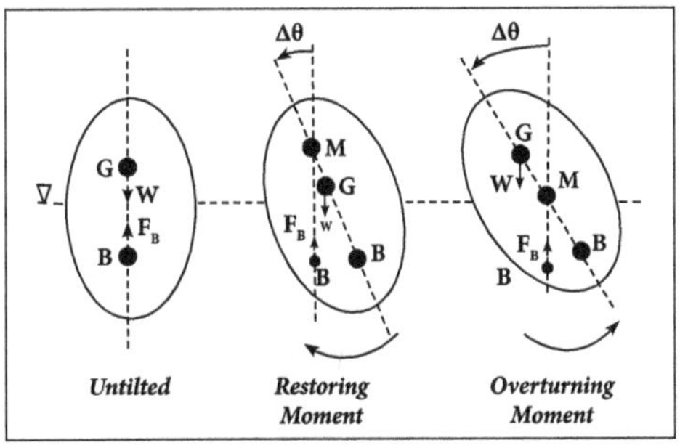

When the body is tilted the centre of buoyancy moves to a new position, B', because the sha pe of the displaced volume changes. A new point, M, may be defined, called the METACENTRE. This is the point where a vertical line drawn upwards from the new centre of buoyancy, B', of the tilted body intersects the line of symmetry of the body. The buoyancy force, FB, now acts through B'.

From the centre diagram in the figure we can see that W and FB give a RESTORING MOMENT that rotates the body back to its untilted position. From the right hand diagram in the figure we can see that W and FB give an overturning moment that rotates the body even further in the tilted direction.

The meta centre, M, lies above the centre of gravity, G, then the body is stable. In other words the METACENTRIC HEIGHT, MG is positive $\left(MG = z_M - z_G > 0\right)$. If the meta centre, M, lies below the centre of gravity, G, then the body is unstable. In other words the metacentric height, MG, is negative $\left(MG < 0\right)$.

The metacentric height, MG, is given by:

$$MG = MB - GB \text{ or } MG = \frac{1}{V_S} - GB$$

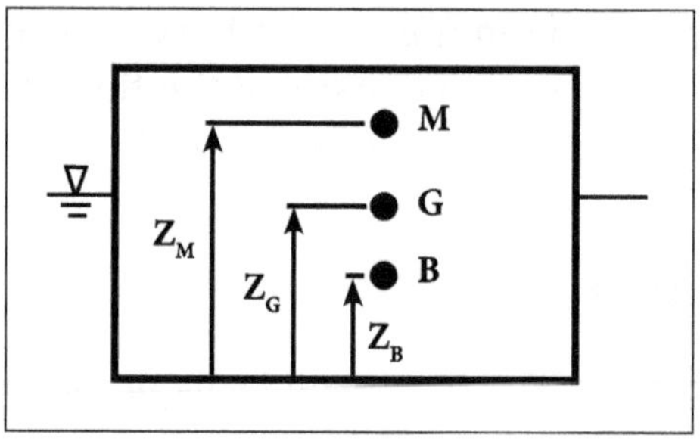

Where, I is the 2nd moment of area of the plan section of the body where it cuts the waterline (this is the solid plane surface if we cut horizontally through a solid body at the water surface lifted the top part up and looked at the bottom of it), VS is the submerged volume and GB is the distance between the centre of gravity and the centre of buoyancy $\left(= z_G - z_B\right)$.

1.5.1 Submerged Bodies: Calculation of Metacenter Height

Meta Center

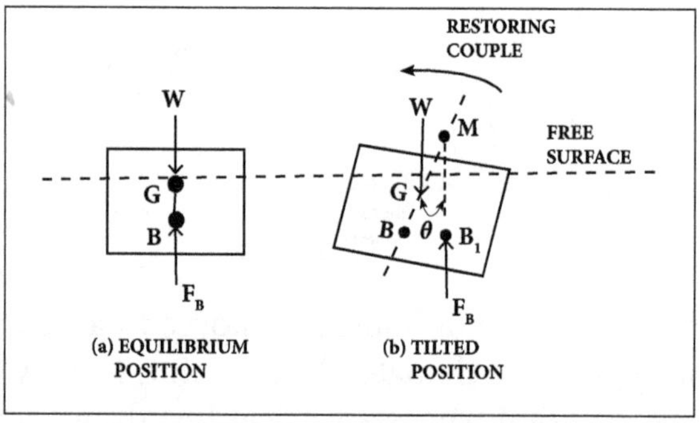

The figure shows the situation after the body has undergone a small angular displacement (θ) with respect to the vertical axis. (G) remains unchanged relative to the body. (B) is the Center of Buoyancy and it moves towards the right to the new position [B1]. The new line of action of the buoyant force through [B1] which is always vertical intersects the axis BG (old vertical line through [B] and [G]) at [M]. For small angles of (θ), point [M] is practically constant and is known as Meta Center.

Meta Center [M] is a point of intersection of the lines of action of Buoyant Force before and after heel. The distance between Center of Gravity and Meta Center (GM) is called Meta-Centric Height. The distance [BM] is known as Meta Centric Radius.

In Figure (b), [M] is above [G], the Restoring Couple acts on the body in its displaced position and tends to turn the body to the original position Floating body is in stable equilibrium.

If [M] were below [G], the couple would be an Over turning Couple and the body would be in Unstable Equilibrium.

If [M] coincides with [G], the body will assume a new position without any further movement and thus will be in Neutral Equilibrium.

For a floating body, stability is determined not simply by the relative positions of [B] and [G]. The stability is determined by the relative positions of [M] and [G]. The distance of the Meta-Center [M] above [G] along the line [BG] is known as the Meta-Centric height (GM).

$$GM = BM - BG$$

$GM > 0$, [M] above [G] ------- Stable Equilibrium

$GM = 0$, [M] coinciding with [G] ------Neutral Equilibrium

$GM < 0$, [M] below [G] ------- Unstable Equilibrium

Determination of Meta-Centric Height

Let,

- w_1 = known weight placed over the center of the vessel as shown in Figure (a) and vessel is floating.
- W=Weight of the vessel including (w1).
- G=Center of gravity of the vessel B=Center of buoyancy of the vessel.

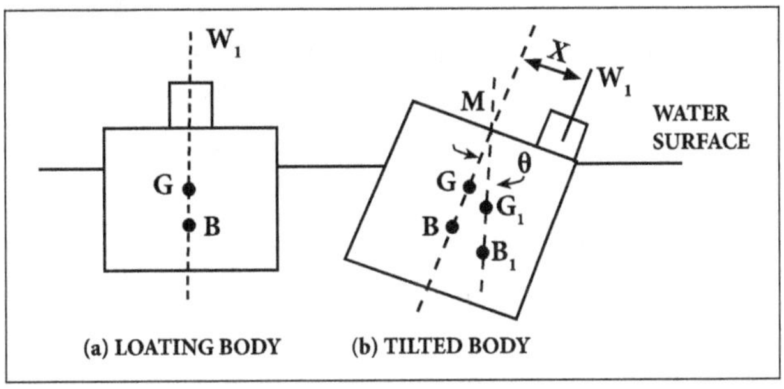

Experimental method for determination of Meta-centric height.

A move weight (w1) across the vessel towards right by a distance (x) as shown in Figure (b). The angle of heel can be measured by means of a plumb line. The new Center of Gravity of the vessel will shift to (G1) and the Center of Buoyancy will change to B_1.

Under equilibrium, the moment caused by the movement of the load (w1) through a distance (x) = Moment caused by the shift of center of gravity from (G) to (G_1).

Moment due to the change of $G = W\left(GG_1\right) = W\left(GM \tan\theta\right)$

Moment due to movement of $w_1 = w_1\left(x\right) = W\left(GM.\tan\theta\right)$

Therefore, $GM = \left[\left(w_1 x\right)/\left(W \tan\theta\right)\right]$

1.5.2 Stability Analysis and Applications

Stability of Un-Constrained Submerged Bodies in a Fluid

When a body is submerged in a liquid, the weight of the body acting through its Center of Gravity should be co-linear with the Buoyancy Force acting through the Center of Buoyancy. If the Body is not homogeneous in its distribution of mass over the entire volume, the location of Center of Gravity (G) does not coincide with the Center of Volume (B).

Depending upon the relative locations of (G) and (B), the submerged body attains different states of equilibrium: Stable, Unstable and Neutral.

1. Stable Equilibrium: (G) is located below (B). A body being given a small angular displacement and then released, returns to its original position by retaining the original vertical axis as vertical because of the restoring couple produced by the action of the Buoyant Force and the weight.

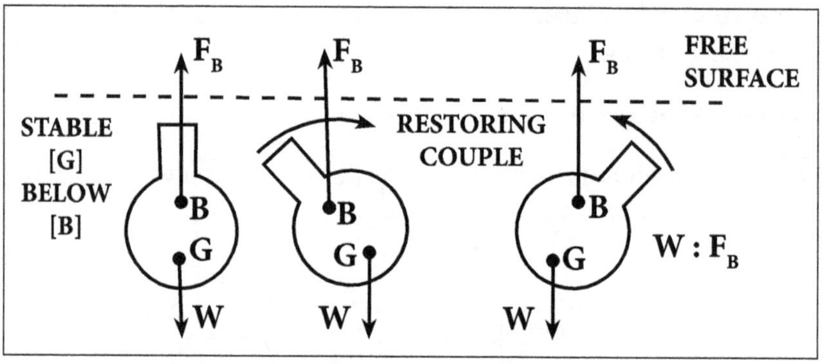

2. Unstable Equilibrium: (G) is located above (B). Any disturbance from the equilibrium position will create a destroying couple that will turn the body away from the original position.

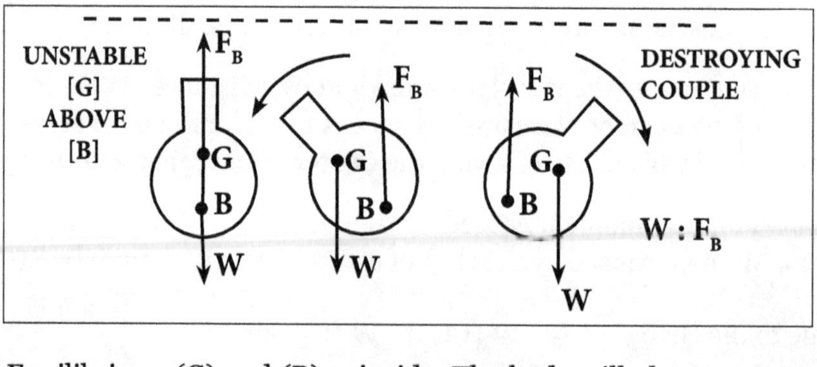

3. Neutral Equilibrium: (G) and (B) coincide. The body will always assume the same position in which it is placed. A body having a small displacement and then released, neither returns to the original position nor increases its displacement. It will simply adapt to the new position.

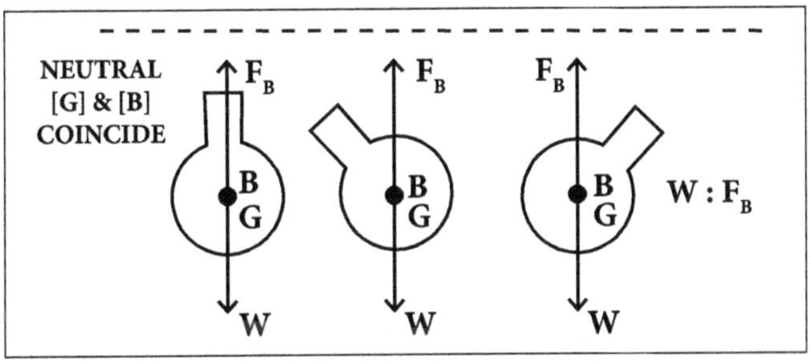

Stability of Floating Bodies

Stable conditions of the floating body can be achieved, under certain conditions even though (G) is above (B). When a floating body undergoes angular displacement about the horizontal position, the shape of the immersed volume changes and so, the Center of Buoyancy moves relative to the body.

2

Fluid Kinematic and Fluid Dynamics

2.1 Fluid Mechanics

Fluid mechanics is defined as the science that deals with the behavior of fluids at rest (fluid statics) or in motion (fluid dynamics) and the interaction of fluids with solids or other fluids at the boundaries.

2.1.1 Flow Types

The types of flow can be classified into:

- Laminar flow.
- Turbulent flow.

A Fluid is substance that can flow, like a Liquid or a Gas. The flow of a fluid can be represented by Streamlines, which are 'fluid elements' that move relative to each other.

Types of Fluid Flow

The fluid flow is classified as:

- Steady and Unsteady flows.
- Uniform and Non – uniform flows.
- Laminar and turbulent flows.
- Compressible and incompressible flows.
- Rotational and irrotational flows.
- One, two and three dimensional flows.

Laminar Flow: The laminar flow is defined as that type of flow in which the fluid particles move along well defined paths or stream line and all the stream lines are straight and parallel. Thus the particles move in laminas or layers gliding over the adjacent layer. This type of flow is also called stream line flow or viscous flow.

Turbulent Flow: Turbulent flow is that type of flow in which the fluid particles move in a zigzag way. Due to the movement of fluid particles in a zigzag way.

Steady Flow: In steady fluid flow, the velocity of the fluid is constant at any point.

Unsteady Flow: When the flow is unsteady, the fluid's velocity can differ between any two points.

Uniform Flow: The flow is defined as uniform flow when in the flow field the velocity and other hydrodynamic parameters do not change from point to point at any instant of time.

Implication,

- For a uniform flow, there will be no spatial distribution of hydrodynamic and other parameters.
- Any hydrodynamic parameter will have a unique value in the entire field, irrespective of whether it changes with time unsteady uniform flow or does not change with time steady uniform flow.
- Thus, steadiness of flow and uniformity of flow does not necessarily go together.

Non-Uniform Flow: When the velocity and other hydrodynamic parameters changes from one point to another the flow is defined as non-uniform.

Important Points,

- A non-uniform flow, the changes with position may be found either in the direction of flow or in directions perpendicular to it.
- Non-uniformity in a direction perpendicular to the flow is always encountered near solid boundaries past which the fluid flows.

Compressible or Incompressible Flow: Fluid flow can be compressible or incompressible, depending on whether we can easily compress the fluid. Liquids are usually nearly impossible to compress, whereas gases are very compressible.

A hydraulic system works only because liquids are incompressible that is, when we increase the pressure in one location in the hydraulic system, the pressure increases to match everywhere in the whole system.

Viscous or Non-Viscous Flow: Viscosity is actually a measure of friction in the fluid. When a fluid flows, the layers of fluid rub against one another and in very viscous fluids, the friction is so great that the layers of flow pull against one other and hamper that flow.

Viscosity usually varies with temperature, because when the molecules of a fluid are moving faster, the molecules can more easily slide over each other. for example, we may notice that it's very thick in the bottle, but the syrup becomes quite runny when it spreads over the warm pancakes and heats up.

Rotational or Irrotational Flow: Fluid flow can be rotational or irrotational. If, as we travel in a closed loop, we add up all the components of the fluid velocity vectors along with the path and the end result is not zero, then the flow is rotational.

To test whether a flow has a rotational component, we can put a small object in the flow and let the flow carry it. If the small object spins, the flow is rotational, if the object doesn't spin, the flow is irrotational.

For example, look at the water flowing in a brook. It eddies around stones, curling around obstacles. At such locations, the water flow has a rotational component.

Some flows that we may think are rotational are actually irrotational. For example, away from the center, a vortex is actually an irrotational flow! We can see this if we look at the water draining from we r bathtub. If we place a small floating object in the flow, it goes around the plug hole, but it does not spin about itself, therefore, the flow is irrotational.

On the other hand, flows that have no apparent rotation can actually be rotational. Take a shear flow, for example. In a shear flow all the fluid is moving in the same direction, but the fluid is moving faster on one side. Suppose the fluid is moving faster on the left than on the right. The fluid isn't moving in a circle at all, but if we place a small floating object in this flow, the flow on the left side of the object is slightly faster so the object begins to spin. The flow is rotational.

One, Two and Three Dimensional Flows: Fluid flow is three-dimensional in nature. This means that the flow parameters like velocity, pressure and so on vary in all the three coordinate directions.

Sometimes simplification is made in the analysis of different fluid flow problems by:

Selecting the appropriate coordinate directions so that appreciable variation of the hydro dynamic parameters take place in only two directions or even in only one.

One-Dimensional Flow: All the flow parameters may be expressed as functions of time and one space coordinate only. The single space coordinate is usually the distance measured along the center line (not necessarily straight) in which the fluid is flowing.

Example: The flow in a pipe is considered one-dimensional when variations of pressure and velocity occur along the length of the pipe, but any variation over the cross-section is assumed negligible.

In reality, flow is never one-dimensional because viscosity causes the velocity to decrease to zero at the solid boundaries.

If however, the non-uniformity of the actual flow is not too great, valuable results may often be obtained from a "one dimensional analysis".

The average values of the flow parameters at any given section (perpendicular to the flow) are assumed to be applied to the entire flow at that section.

Two-Dimensional Flow: All the flow parameters are functions of time and two space coordinates (say x and y).

No variation in z direction. The same streamline patterns are found in all planes perpendicular to z direction at any instant.

Three-Dimensional Flow: The hydrodynamic parameters are functions of three space coordinates and time.

Table: Comparison of laminar and turbulent flow.

S. No.	Laminar Flow	Turbulent Flow
1.	Laminar flow is one in which the fluid particles are more within a specified layer, with one layer of fluid sliding smoothly over an adjacent layer.	Turbulent flow is one in which the fluid particles move in an entirely haphazardard or erratic manner.
2.	The fluid particles move in well-defined paths.	Fluid particles move in a un predictable path.
3.	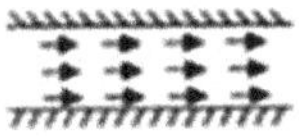 	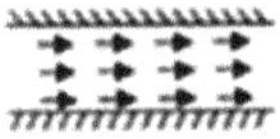 
4.	This type of flow occurs in smooth pipes when the velocity of flow is low.	This type of flow occurs in rivers, canals, stream and etc.

Examples of Laminar Flow:

- Smoke at outlet of Chimney.
- High viscous liquid flow in smaller diameter Pipe with less velocity.
- Flow in the smooth pipes.

Examples of Turbulent Flow:

- Blood flow in arteries.
- Oil transport in pipelines, lava flow.
- The flow through pumps and turbines.

Reynolds Number

Reynolds Number is defined as the ratio of inertial force of a flowing fluid and the viscous force of the fluid.

$$\text{Reynolds Number for pipe flow, } R_e = \frac{\rho V d}{\mu}$$

Where,

- ρ = Density of fluid.
- V = Velocity.
- d = diameter of the pipe
- μ = Viscosity of fluid.
- For laminar flow $R_e < 2000$.
- For turbulent flow $R_e > 4000$.

Significance:

- It signifies the relative predominance of inertia to viscous forces.
- It is very useful in determining whether the flow is laminar or turbulent.

2.1.2 Equation of Continuity for 1D Flow

When a fluid is motion, it must move in such a way that mass is conserved. Consider the steady flow of fluid through a duct (the inlet and outlet flows do not vary with time). The inflow and outflow are one-dimensional, so that the velocity V and density ρ are constant over the area A.

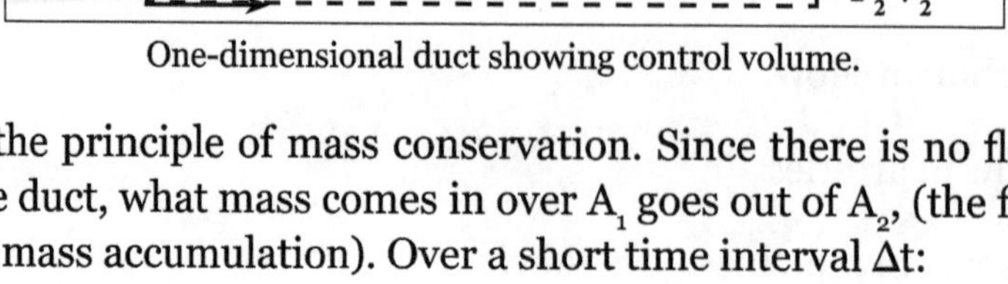

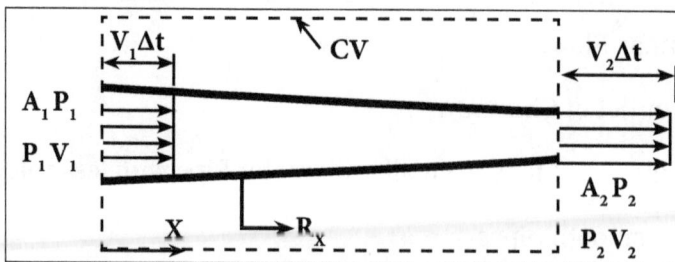

One-dimensional duct showing control volume.

Now we apply the principle of mass conservation. Since there is no flow through the side walls of the duct, what mass comes in over A_1 goes out of A_2, (the flow is steady so that there is no mass accumulation). Over a short time interval Δt:

$$\text{Volume flow in over } A_1 = A_1 V_1 \Delta t$$

$$\text{Volume flow in over } A_2 = A_2 V_2 \Delta t$$

Therefore,

$$\text{Mass in over } A = \rho A_1 V_1 \Delta t$$

$$\text{Mass out over } A = \rho A_2 V_2 \Delta t$$

So,

$$\rho A_1 V_1 = \rho A_2 V_2$$

This is a statement of the principle of mass conservation for a steady, one-dimensional flow with one inlet and one outlet. This equation is called the continuity equation for steady one-dimensional flow. For a steady flow through a control volume with many inlets and outlets, the net mass flow must be zero, where inflows are negative and outflows are positive.

2.2 Circulation and Vorticity

Vorticity is defined as the curl of the velocity field and is hence a measure of local rotation of the fluid. This definition makes it a vector quantity. Circulation on the other hand, is a scalar quantity defined as the line integral of the velocity field along a closed contour. Using Stoke's theorem, the line integral of the velocity field along the closed path can be expressed as a surface integral of the curl of the velocity field normal to an arbitrary area bounded by the path. Hence circulation can be referred to as flux of vorticity. It can also be said that that vorticity at a point is essentially circulation per unit area. The last two statements characterize the two quantities, vorticity and circulation, as microscopic and macroscopic respectively. Both these quantities are essentially a measure of the rotation of the fluid flow.

As far as the physical meaning is concerned, circulation can be through as the amount of 'push' one feels while moving along a closed boundary or path. Vorticity however has nothing to do with a path; it is defined at a point and would indicate the rotation in the flow field at that point. So, if an infinitesimal paddle wheel is imagined in the flow, it would rotate due to non-zero vorticity.

2.2.1 Streamline, Stream Tubes, Path Line and Streak Lines

Streamline

A streamline is defined as a line which is everywhere parallel to the local velocity vector:

$$\vec{V}(x,y,z,t) = u\hat{i} + v\hat{j} + w\hat{k}$$

$$d\vec{s} = dx\,\hat{i} + dy\,\hat{j} + dz\,\hat{k}$$

As an infinitesimal arc length vector along the streamline. Since this is parallel to $\vec{V}$, we must have:

$$d\vec{s} \times \vec{V} = 0$$

$$(w\,dy - v\,dz)\hat{i} + (u\,dz - w\,dx)\hat{j} + (v\,dx - u\,dy)\hat{k} = 0$$

Separately setting both components to zero gives three differential equations which define the streamline. The three velocity components u, v, w, must be given as functions of x, y, z before these equations can be integrated. To set the constants of integration, it is sufficient to specify some point x_0, y_0, z_0 through which the streamline passes.

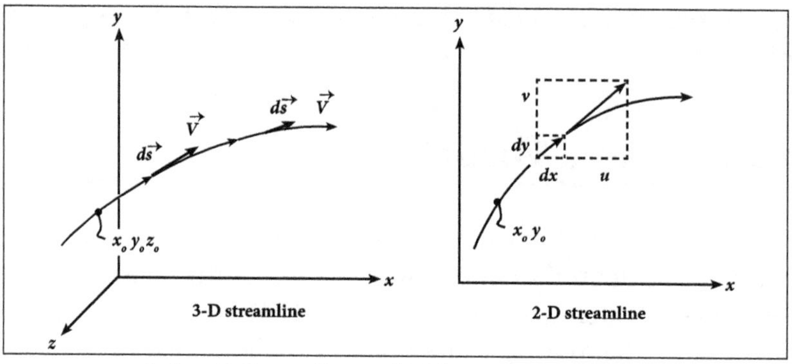

Left: 3-D Streamline and Right: 2-D Streamline.

Stream Tubes

It considers a set of x_0, y_0, z_0 points arranged in a closed loop. The streamlines passing through all these points form the surface of a stream tube. Because there is no flow across the surface, each cross-section of the stream tube carries the same mass flow. So the stream tube is equivalent to a channel flow embedded in the rest of the flow field.

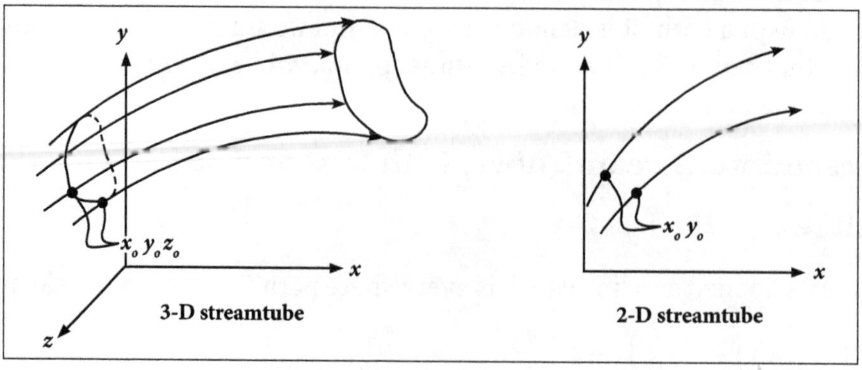

Left: 3-D streamtube and Right: 2-D streamtube.

In 2-D a stream tube is defined by two streamlines passing through two specified x_o, y_o points. The flow between these two streamlines carries the same mass flow or span at each cross-section and can be considered as a 2-D channel flow embedded in the rest of the flow field.

Path Lines

The path line of a fluid element A is simply the path it takes through space as a function of time. An example of a path line is the flight path taken by one of smoke which is carried by the steady or unsteady wind. This path is fully described by the three position functions $x_A(t)$, $y_A(t)$, $z_A(t)$, which can be computed by integrating the three velocity field components u(x, y, z, t), v(x, y, z, t), w(x, y, z, t) along the path. The integration is started at time to, from the element's initial position x_o, y_o, z_o (e.g. the smoke release point) and proceeds to some later time 't'.

$$x_A(t)=x_o+\int_{t_o}^{t} u\left(x_A(\tau),y_A(\tau),z_A(\tau),\tau\right)d\tau$$

$$y_A(t)=y_o+\int_{t_o}^{t} v\left(x_A(\tau),y_A(\tau),z_A(\tau),\tau\right)d\tau$$

$$z_A(t)=z_o+\int_{t_o}^{t} w\left(x_A(\tau),y_A(\tau),z_A(\tau),\tau\right)d\tau$$

The dummy variable of integration τ runs from to to t.

Streak Lines

A streak line is associated with a particular point P in space which has the fluid moving past it. All points which pass through this point are said to form the streak line of point P. An example of a streak line is the continuous line of smoke emitted by a chimney at point P, which will have some curved shape if the wind has a time varying direction.

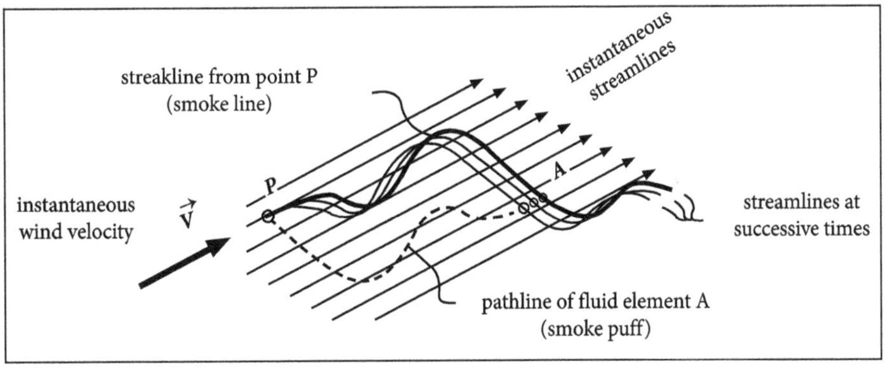

2.3 Stream Function and Velocity Potential Function

The differential relations for fluid particle can be written for conservation of mass, momentum and energy. In addition, there are two state relations for thermodynamic properties. They are:

$$\text{Continuity:}\frac{\partial \rho}{\partial t}+\nabla.\left(\rho\vec{V}\right)=0$$

$$\text{Momentum: }\rho\frac{d\vec{V}}{dt}=\rho\vec{g}-\nabla p+\nabla.\tau_{ij} \qquad ...(1)$$

$$\text{Energy:}\rho\frac{\partial e}{dt}+p\left(\nabla.\vec{V}\right)=\nabla.\left(k\nabla T\right)+\Phi$$

$$\text{Thermodynamic state relation:}\rho=\rho\left(p,T\right);\; e=e\left(p,T\right)$$

The idea of introducing stream function works only if the continuity equation is reduced to two terms. There are four-terms in the continuity equation that one can get by expanding the vector equation (1).

$$\frac{\partial \rho}{\partial t}+\frac{\partial(\rho u)}{\partial x}+\frac{\partial(\rho v)}{\partial y}+\frac{\partial(\rho w)}{\partial z}=0 \qquad ...(2)$$

For a steady, incompressible, plane, two-dimensional flow, this equation reduces to:

$$\frac{\partial u}{\partial x}+\frac{\partial v}{\partial y}=0 \qquad ...(3)$$

Here, the striking idea of stream function works that will eliminate two velocity components u and v into a single variable shown in figure. So, the stream function $\left\{\Psi(x,y)\right\}$ relates to the velocity components in such a way that continuity equation (3) is satisfied.

$$u=\frac{\partial \Psi}{\partial y};v=-\frac{\partial \Psi}{\partial x}$$

Or,

$$\vec{V}=\frac{\partial \Psi}{\partial y}\hat{i}-\frac{\partial \Psi}{\partial x}\hat{j} \qquad ...(4)$$

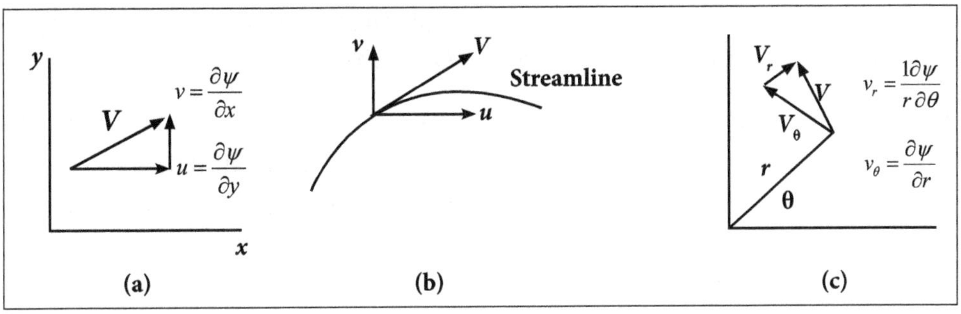

(A) Velocity components along a streamline.

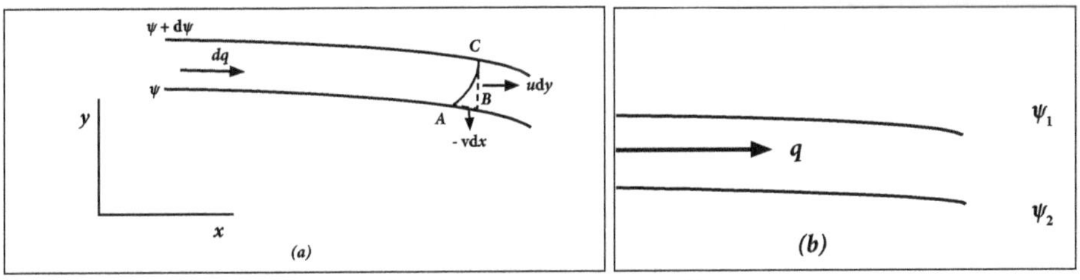

(B) Flow between two streamlines.

The following important points can be noted for stream functions:

1. The lines along which Ψ is constant are called as stream lines. In a flow field, the tangent drawn at every point along a streamline shows the direction of velocity. So, the slope at any point along a streamline is given by:

$$\frac{dy}{dx} = \frac{v}{u} \qquad ...(5)$$

Referring to the figure, (B) -a, if we move from one point (x, y) to a nearby point $(x+dx, y+dy)$, then the corresponding change in the value of stream function is d Ψ which is given by:

$$d\Psi = \frac{\partial \Psi}{\partial x}dx + \frac{\partial \Psi}{\partial y}dy = -v\,dx + u\,dy \qquad ...(6)$$

Along a line of constant Ψ,

$$d\Psi = -v\,dx + u\,dy = 0$$

Or,

$$\frac{dy}{dx} = \frac{v}{u} \qquad ...(7)$$

2. The Equation (5) is same as that of Equation (7). Hence, it is the defining equation for the streamline. Thus, infinite number streamlines can be drawn with constant Ψ.

This family of streamlines will be useful in visualizing the flow patterns. It may also be noted that streamlines are always parallel to each other.

3. The numerical constant associated to Ψ, represents the volume rate of flow. Consider two closely spaced stream lines Ψ and $(\Psi + d\Psi)$ as shown in figure (B) -a. Let $d\dot{q}$ represents the volume rate of flow per unit width perpendicular to x-y plane, passing between the streamlines. At any arbitrary surface AC, this volume flow must be equal to net outflow through surfaces AB and BC. Thus,

$$d\dot{q} = -v\,dx + u\,dy = \frac{\partial \Psi}{\partial x}dx + \frac{\partial \Psi}{\partial y}dy = d\Psi$$

Or,

$$d\dot{q} = d\Psi \qquad ...(8)$$

Hence, the volume flow rate $(\dot{q})$ can be determined by integrating Equation (8). Between streamlines Ψ_1 and Ψ_2 as follows:

$$\dot{q} = \int_{\Psi_1}^{\Psi_2} d\Psi = \Psi_2 - \Psi_1 \qquad ...(9)$$

So, the change in the value of stream function is equal to volume rate of flow. If the upper streamline Ψ_2 has a value greater than the lower one Ψ_1, then the volume flow rate is positive i.e. flow takes place from left to right (Figure (B) -b).

4. In cylindrical coordinates, the continuity equation for a steady, incompressible, plane, two-dimensional flow, reduces to:

$$\frac{1}{r}\frac{\partial (r v_r)}{\partial r} + \frac{1}{r}\frac{\partial v_\theta}{\partial \theta} = 0 \qquad ...(10)$$

The respective velocity components v_γ and v_θ are shown in Figure (A) -c. The stream function $\{\Psi(\gamma, \theta)\}$ that satisfies equation (10), can then be defined as:

$$v_r = \frac{1}{r}\frac{\partial \Psi}{\partial \theta}; v_\theta = -\frac{\partial \Psi}{\partial r} \qquad ...(11)$$

4. In a steady, plane compressible flow, the stream function can be defined by including the density of the fluid. But, the change in the stream function is equal to mass flow rate $(\dot{m})$.

$$\rho u = \frac{\partial \Psi}{\partial y}; \rho v = -\frac{\partial \Psi}{\partial x}$$

$$d\dot{m} = -\rho v\, dx + \rho u\, dy = \frac{\partial \Psi}{\partial x} dx + \frac{\partial \Psi}{\partial y} dy = d\Psi$$

Or,

$$d\dot{m} = d\Psi \qquad ...(12)$$

5. One important application in a two-dimensional plane is the in viscid and irrotational flow where, there is no velocity gradient and $\omega_z = 0$. Then, the vorticity vector becomes:

$$\zeta = 2\omega_z \hat{k} = \left(\frac{\partial v}{\partial x} - \frac{\partial u}{\partial y} \right) \hat{k} = 0$$

$$\text{or,} \left[\frac{\partial}{\partial x} \left(-\frac{\partial \psi}{\partial x} \right) - \frac{\partial}{\partial y} \left(\frac{\partial \psi}{\partial y} \right) \right] \hat{k} = 0$$

$$\text{or,} \frac{\partial^2 \psi}{\partial x^2} - \frac{\partial^2 \psi}{\partial y^2} = 0$$

$$\text{or,} \nabla^2 \psi = 0 \qquad ...(13)$$

This is a second order equation and is quite popular in mathematics and is known as Laplace equation in a two-dimensional plane.

Velocity Potential

An irrotational flow is defined as the flow where the vorticity is zero at every point. It gives rise to a scalar function (ϕ) which is similar and complementary to the stream function (ψ). Let us consider the equations of irrotational flow and scalar function (ϕ). In an irrotational flow, there is no vorticity $(\vec{\xi})$:

$$\vec{\xi} = \nabla \times \vec{V} = 0 \qquad ...(14)$$

Now, take the vector identity of the scalar function (ϕ):

$$\nabla \times (\nabla \phi) = 0 \qquad ...(15)$$

i.e. a vector with zero curl must be the gradient of a scalar function or, curl of the gradient of a scalar function is identically zero. Comparing, Equation (14) and (15), we see that:

$$\vec{V} = \nabla \phi \qquad ...(16)$$

Here, ϕ is called as velocity potential function and its gradient gives rise to velocity vector. The knowledge ϕ immediately gives the velocity components. In Cartesian coordinates, the velocity potential function can be defined as, $\phi=\phi(x,y,z)$ so that Equation (16) can be written as:

$$u\hat{i}+v\hat{j}+w\hat{k}=\frac{\partial\phi}{\partial x}\hat{i}+\frac{\partial\phi}{\partial y}\hat{j}+\frac{\partial\phi}{\partial z}\hat{k} \quad ...(17)$$

So, the velocity components can be written as:

$$u=\frac{\partial\phi}{\partial x};\ v=\frac{\partial\phi}{\partial y};\ w=\frac{\partial\phi}{\partial z} \quad ...(18)$$

In cylindrical coordinates, if $\phi=\phi(r, \theta, z)$, then:

$$V_r=\frac{\partial\phi}{\partial r};V_\theta=\frac{\partial\phi}{\partial\theta};V_z=\frac{\partial\phi}{\partial z} \quad ...(19)$$

Further, if the flow is incompressible i.e. ρ=constant and $(\partial\rho/\partial t)=0$, then continuity equation can be written as:

$$\frac{\partial\rho}{\partial t}+\nabla.(\rho\vec{V})=0$$

$$or, \rho(\rho.\vec{V})=0$$

$$or, \nabla.\vec{V}=0 \quad ...(20)$$

Therefore, for a flow which is incompressible and irrotational, Equation (16) and (20) can be combined to yield a second order Laplace equation in a three-dimensional plane.

$$\nabla.(\nabla\phi)=0$$

$$or, \nabla^2\phi=0$$

$$or, \frac{\partial^2\phi}{\partial x^2}+\frac{\partial^2\phi}{\partial y^2}+\frac{\partial^2\phi}{\partial z^2}=0 \quad ...(21)$$

Thus, any irrotational incompressible flow has a velocity potential and stream function that both satisfy Laplace equation. Conversely any solution of Laplace equation represents both velocity potential and stream function (two-dimensional) for an irrotational, incompressible flow.

An irrotational flow allows a velocity potential to be defined and leads to simplification of fundamental equations. Instead of dealing with the velocity components u, v and w as unknowns, one can deal with only one parameter ϕ, for a given problem. Since, the irrotational flows are best described by velocity potential, such flows are called as potential flows. In these flows, the lines with constant ϕ, is known as equipotential lines. In addition a line drawn in space such that $\nabla\phi$ is the tangent at every point is defined as a gradient line and thus can be called as streamline.

Stream Function vs. Velocity Potential

The velocity potential is analogous to stream function in a sense that the derivatives of both ϕ and ψ yield the flow field velocities. However, there are distinct differences between ϕ and ψ:

- The flow field velocities are obtained by differentiating ϕ in the same direction as the velocities, whereas, ψ is differentiated normal to the velocity direction.
- The velocity potential is defined for irrotational flows only. In contrast, stream function can be used in either rotational or irrotational flows.
- The velocity potential applies to three-dimensional flows, whereas the stream function is defined for two dimensional flows only.

It is seen that the stream lines are defined as lines of constant ψ which are same as gradient lines and perpendicular to lines of constant ϕ. So, the equipotential lines and stream lines are mutually perpendicular. In order to illustrate the results more clearly, let us consider a two-dimensional, irrotational, incompressible flow in Cartesian coordinates.

For a streamline, $\psi(x,y)=\text{constant}$ and the differential of ψ is zero.

$$d\psi=\frac{\partial\psi}{\partial x}dx+\frac{\partial\psi}{\partial y}dy=0$$

$$\text{or}, d\psi=-v\,dx+u\,dy=0$$

$$\text{or}, \left(\frac{dy}{dx}\right)_{\psi=\text{constant}}=\frac{v}{u} \qquad ...(22)$$

Similarly, for an equipotential line, $\phi(x,y)=\text{constant}$ and the differential of ϕ is zero.

$$d\phi=\frac{\partial\phi}{\partial x}dx+\frac{\partial\phi}{\partial y}dy=0$$

$$\text{or, } d\phi = u\,dx + v\,dy = 0$$

$$\text{or,} \left(\frac{dy}{dx}\right)_{\phi=\text{constant}} = -\frac{u}{v} \qquad ...(23)$$

Combining Equation (22) and (23), we can write:

$$\left(\frac{dy}{dx}\right)_{\psi=\text{constant}} = -\frac{1}{\left(\frac{dy}{dx}\right)_{\phi-\text{constant}}} \qquad ...(24)$$

Hence, the streamlines and equipotential lines are mutually perpendicular.

2.4 Euler's and Bernoulli's Equations for Flow along a Stream Line

Euler's Equation

Euler equation of motion is given by:

$$\frac{\partial p}{\rho} + g\,dz + v\,dv = 0$$

Where,

- $P \rightarrow$ Pressure of the fluid.
- $\rho \rightarrow$ Density of the fluid.
- $g \rightarrow$ Acceleration due to gravity.
- $v \rightarrow$ Mean velocity of fluid.

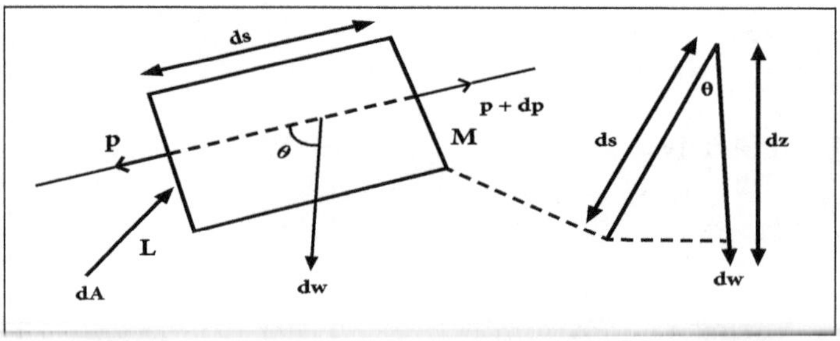

Net Force:

$$PdA - (dp)dA = 0$$

$$PdA - PdA - dp\,dA = 0$$

$$-dp\,dA = 0$$

Weight of the fluid:

$$d\omega = .dA.ds$$

$$\cos\theta = \frac{dz}{ds}$$

$$d\omega = -egdAds.\frac{dz}{dS}$$

$$d\omega = -egdAdz$$

Resultant force for direction of live flow:

$$\text{Force}\,(f) = -\ dp\ dA\ -\ \rho\ g\ dA\ dz$$

According to Newton's Law:

$$F = ma$$

$$\text{Mass}\,(m) = \ \rho.dA\ ds$$

$$d = \frac{dw}{ds}$$

$$d = \frac{dw}{ds} \times \frac{ds}{dt}$$

$$d = \frac{dw}{ds} \times v$$

$$-dpdA - egdAdz = edAds \times \frac{dw}{dS} \times V$$

Divide edA:

$$-\frac{dp}{e} - 2dz = V.dv$$

$$\frac{dp}{e} + 2dz + vdw = 0$$

From bernoullis equation:

$$\frac{1}{e}\int dp+g\int dz+\int V\,dv=0$$

$$\frac{p}{e}+2Z+\frac{V^L}{2}=0$$

Divide g:

$$\frac{p}{e}+\frac{V^L}{2g}+z=0$$

$$\frac{P_1}{eg}+\frac{V_1^2}{2g}+Z_1=\frac{P_2}{eg}+\frac{V_2^L}{2g}+z \quad (\text{for two sections})$$

Assumptions Made in Euler Equation

- The fluid is non-viscous (i.e., the frictional losses are zero).
- The fluid is homogeneous and incompressible (i.e., mass density of the fluid is constant).
- The flow is continuous, steady and along the streamline.
- The velocity of the flow is uniform over the section.
- No energy or force (except gravity and pressure forces) is involved in the flow.

2.5 Closed Conduit Flow: Reynold's Experiment and Darcy Weisbach Equation

2.5.1 Reynolds's Experiment

The help of a simple experiment conducted in a glass tube with Coloured liquids Reynolds determined the state of laminar and turbulent flow. He observed in the glass tube that at low velocity, Coloured water moved in a straight path. With increasing velocity a observed the flow of Coloured water to be somewhat irregular. With further increase in the velocity, the fluctuation of the filament became more intense and the dye diffused over the entire cross-section of the tube.

This is the turbulent state of the flow. The velocity at which the flow begins to enter from the laminar to the turbulent state is called critical velocity. Reynolds observed that the occurrence of these states of flow is governed by relative magnitudes of inertia and viscous force. He related the inertial force to viscous force by the dimensionless number R.

$$R_e = \frac{\text{Inertia force}}{\text{Viscous force}} = \frac{F_i}{F_v}$$

According to Newton's second law of motion, this inertial force F1 is given by:

$$F_i = \text{mass} \times \text{acceleration}$$

$$= \rho(\text{volume}) \times \text{acceleration}$$

$$= \rho L^3 \cdot \left(\frac{L}{T^2}\right)$$

$$= \rho L^2 \left(\frac{L^2}{T^2}\right)$$

$$= \rho L^2 V^2 \qquad \because \frac{L}{T} = V$$

Similarly, viscous force $F_v = \text{shear stress} \times \text{area}$

$$= \mu \frac{\partial v}{\partial y} L^2$$

$$= \mu \left(\frac{V}{L}\right) L^2$$

$$= \mu VL$$

$$\left. \begin{aligned} R_e = \frac{\rho L^2 V^2}{\mu VL} &= \frac{\rho VL}{\mu} \\ &= \frac{VL}{\frac{\mu}{\rho}} \\ &= \frac{VL}{v} \end{aligned} \right\}$$

Where,

- ρ - Density.

- μ - Dynamic viscosity.
- L - Length of the pipe.
- D - Diameter of the pipe.
- $\frac{\mu}{\rho} = v$ is called kinetic viscosity.
- For laminar flow $R_e < 2000$.
- For turbulent flow $R_e > 4000$.

2.5.2 Darcy's Equation

The Darcy equation is given by:

$$h_f = \frac{4fLV^2}{d \times 2g}$$

Where,

- $h_f \rightarrow$ Loss of head due to friction.
- $f \rightarrow$ Co-efficient of friction.
- $L \rightarrow$ Length of the pipe.
- $V \rightarrow$ Velocity of flow.
- $d \rightarrow$ Diameter of pipe.
- $g \rightarrow$ Acceleration due to gravity.

2.6 Momentum Equation and Its Applications

Equation of Continuity and the Momentum Equation

General expressions can be utilized to determine velocity distributions for flow systems. This method is better than developing formulations peculiar to the specific problem at hand. The general momentum equation is also called the equation of motion or the Navier-Stoke's equation.

The equation of continuity is developed simply by applying the law of conservation of mass to a small volume element within a flowing fluid. The momentum equation (the equation of motion or the Navier Stoke's equation is an extension of previously written momentum balance.

Due to the conservation of mass:

(Rate of mass accumulation) = (Rate of mass in) – (Rate of mass out)

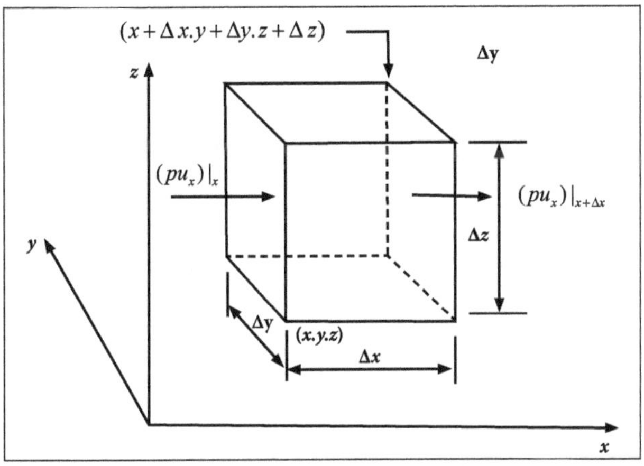

Volume element fixed in space with fluid flow.

Consider a stationary volume within a fluid moving with a velocity having the components v_x, v_y and v_z.

The volume flow rate of fluid (VFR) in or out across the face = the velocity x the cross-sectional area

The rate of mass in or out through the face = (VFR) x density of fluid

$$\Delta x\,\Delta y\,\Delta z\frac{\partial \rho}{\partial t}=\Delta y\,\Delta z\left[\rho v_x|_x-\rho v_x|_{x+\Delta x}\right]$$

$$+\Delta x\,\Delta z\left[\rho v_y|_y-\rho v_y|_{y+\Delta y}\right]+\Delta x\,\Delta y\left[\rho v_z|_z-\rho v_z|_{z+\Delta z}\right]$$

Then dividing through by $\Delta x, \Delta y, \Delta z$ and taking the limit as these dimensions approach zero, we get the equation of continuity:

$$\frac{\partial \rho}{\partial t}=-\left[\frac{\partial}{\partial x}\rho v_x+\frac{\partial}{\partial y}\rho v_y+\frac{\partial}{\partial z}\rho v_z\right]$$

If the fluid density is constant then the continuity equation reduces to given by equation:

$$0=\left[\frac{\partial v_x}{\partial x}+\frac{\partial v_y}{\partial y}+\frac{\partial v_z}{\partial z}\right]$$

Or, in vector notation $\nabla.v=0$.

Rectangular coordinates (x, y, z):

$$\frac{\partial \rho}{\partial t}+\frac{\partial}{\partial x}(\rho v_x)+\frac{\partial}{\partial y}(\rho v_y)+\frac{\partial}{\partial z}(\rho v_z)=0$$

Cylinderical coordinates (r,θ,z):

$$\frac{\partial \rho}{\partial t}+\frac{1}{r}\frac{\partial}{\partial r}(\rho r v_r)+\frac{1}{r}\frac{\partial}{\partial \theta}(\rho v_\theta)+\frac{\partial}{\partial z}(\rho v_z)=0$$

Spherical coordinates (r,θ,ϕ):

$$\frac{\partial \rho}{\partial t}+\frac{1}{r^2}\frac{\partial}{\partial r}(\rho r^2 v_r)+\frac{1}{r\sin\theta}\frac{\partial}{\partial \theta}(\rho v_\theta \sin\theta)+\frac{1}{r\sin\theta}\frac{\partial}{\partial \phi}(\rho v_\phi)=0$$

The Momentum Equation

The momentum balance equation is extended to include unsteady-state systems:

(Rate of momentum accumulation) = (Rate of momentum in) – (Rate of momentum out) + (Sum of forces acting on the system)

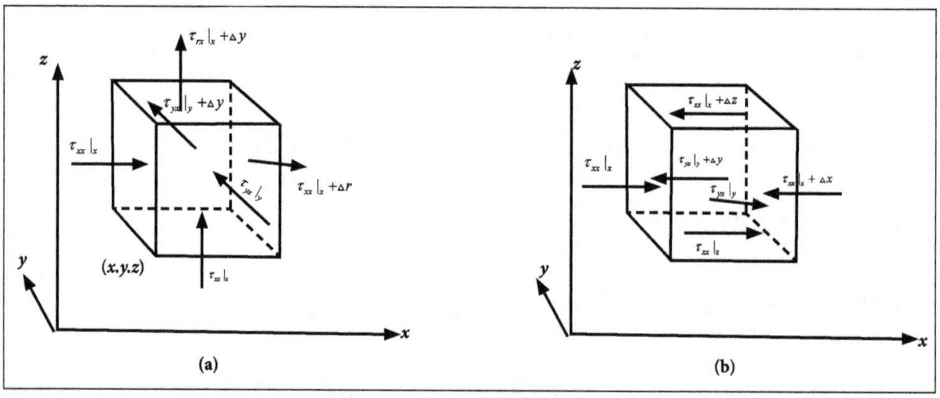

Momentum transport in x component due to viscosity (a) Direction of viscous medium transport (b) Direction of viscous momentum transport.

The x- component of the momentum equation:

$$\frac{\partial}{\partial t}\rho v_x=-\left[\frac{\partial}{\partial x}\rho v_x v_x+\frac{\partial}{\partial y}\rho v_y v_x+\frac{\partial}{\partial z}\rho v_z v_x\right]$$

$$-\left[\frac{\partial}{\partial x}\tau_{xx}+\frac{\partial}{\partial y}\tau_{yx}+\frac{\partial}{\partial z}\tau_{zx}\right]-\frac{\partial}{\partial x}+\rho g_x$$

To describe the general case, all three components (x, y and z) are needed.

In terms of τ:

$$x-\text{component}:\rho\left(\frac{\partial v_x}{\partial t}+v_x\frac{\partial v_x}{\partial x}+v_y\frac{\partial v_x}{\partial y}+v_z\frac{\partial v_x}{\partial z}\right)=-\frac{\partial P}{\partial x}-\left(\frac{\partial \tau_{xx}}{\partial x}+\frac{\partial \tau_{yx}}{\partial y}+\frac{\partial \tau_{zx}}{\partial z}\right)+\rho g_x$$

$$y-\text{component}:\rho\left(\frac{\partial v_y}{\partial t}+v_x\frac{\partial v_y}{\partial x}+v_y\frac{\partial v_y}{\partial y}+v_z\frac{\partial v_x}{\partial z}\right)=-\frac{\partial P}{\partial y}-\left(\frac{\partial \tau_{xy}}{\partial x}+\frac{\partial \tau_{yy}}{\partial y}+\frac{\partial \tau_{zy}}{\partial z}\right)+\rho g_y$$

$$z-\text{component}:\rho\left(\frac{\partial v_z}{\partial t}+v_x\frac{\partial v_z}{\partial x}+v_y\frac{\partial v_z}{\partial y}+v_z\frac{\partial v_z}{\partial z}\right)=-\frac{\partial P}{\partial z}-\left(\frac{\partial \tau_{xz}}{\partial x}+\frac{\partial \tau_{yz}}{\partial y}+\frac{\partial \tau_{zz}}{\partial z}\right)+\rho g_z$$

In terms of velocity gradients for a Newtonian fluid with constant ρ and η:

$$x-\text{component}:\rho\left(\frac{\partial v_x}{\partial t}+v_x\frac{\partial v_x}{\partial x}+v_y\frac{\partial v_x}{\partial y}+v_z\frac{\partial v_x}{\partial z}\right)=-\frac{\partial P}{\partial x}+\eta\left(\frac{\partial^2 v_x}{\partial x^2}+\frac{\partial^2 v_x}{\partial y^2}+\frac{\partial^2 v_x}{\partial z^2}\right)+\rho g_x$$

$$y-\text{component}:\rho\left(\frac{\partial v_y}{\partial t}+v_x\frac{\partial v_y}{\partial x}+v_y\frac{\partial v_y}{\partial y}+v_z\frac{\partial v_y}{\partial z}\right)=-\frac{\partial P}{\partial y}+\eta\left(\frac{\partial^2 v_y}{\partial x^2}+\frac{\partial^2 v_y}{\partial y^2}+\frac{\partial^2 v_y}{\partial z^2}\right)+\rho g_y$$

$$z-\text{component}:\rho\left(\frac{\partial v_z}{\partial t}+v_x\frac{\partial v_z}{\partial x}+v_y\frac{\partial v_z}{\partial y}+v_z\frac{\partial v_z}{\partial z}\right)=-\frac{\partial P}{\partial z}+\eta\left(\frac{\partial^2 v_z}{\partial x^2}+\frac{\partial^2 v_z}{\partial y^2}+\frac{\partial^2 v_z}{\partial z^2}\right)+\rho g_z$$

2.7 Minor Losses, Total Energy Line and Hydraulic Gradient Line

2.7.1 Minor Loss

The minor losses of pipe includes:

- Loss of energy due to sudden enlargement.
- Loss of energy due to sudden contraction.
- Loss of energy due to an obstruction in a pipe.
- Loss of energy due to bend in the pipe.
- Loss of energy at the entrance to a pipe.

- Loss of energy at the exit of a pipe.
- Loss of energy in pipe fittings.

1. Loss of energy due to sudden enlargement:

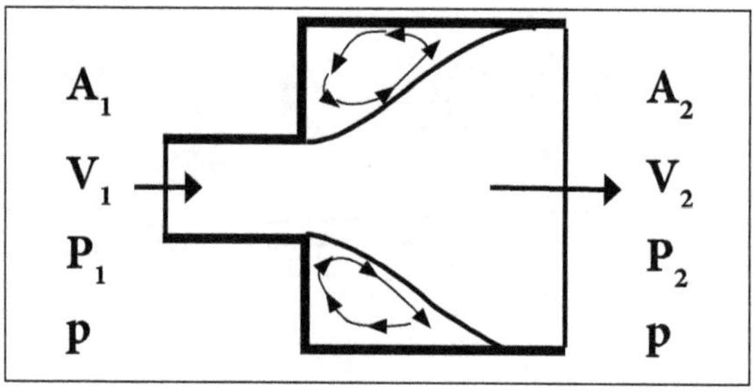

Due to sudden enlargement of pipe the liquid flowing from smaller pipe will not be able to follow the abrupt change of boundary. So the flow separates from the boundary. Because of this separation, the turbulent eddies are formed at these region which result in the loss of energy.

$$h_e = \frac{(V_1 - V_2)^2}{2g}$$

2. Loss of energy due to sudden Contraction:

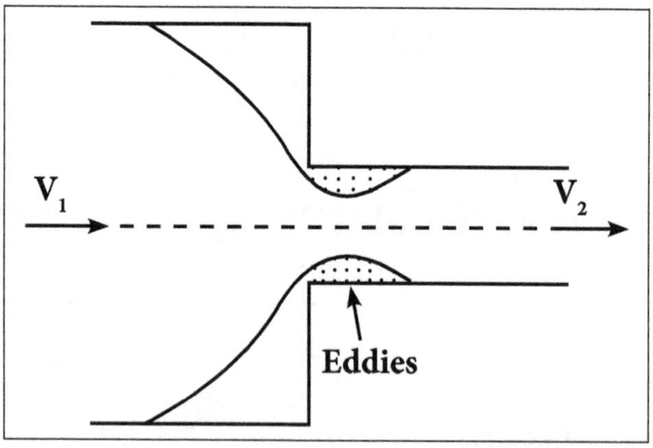

Due to sudden contraction, the stream converges to a minimum cross section called the vena contract and then expands to fill the down steam pipe. In between the vena contract and the wall of the pipe, a lot of eddies are formed as shown in above the figure. These eddies cause a considerable dissipation of energy.

Head loss due to sudden contraction:

$$h_c = \frac{V_2^2}{2g}\left[\frac{1}{C_c} - 1\right]^2$$

Where, Cc is the coefficient of contraction.

3. Loss of energy at inlet (h_i):

This type of loss is similar to the loss due to contraction.

$$h_i = 0.5\frac{V^2}{2g}$$

4. Loss of energy at the exit from a pipe (h_o):

The velocity of liquid at outlet of the pipe is dissipated either in the form of free jet or it is lost in the reservoir.

$$h_o = \frac{V^2}{2g}$$

5. Loss of energy due to an obstruction in a pipe (h_{obs}):

The loss of energy due to an obstruction in a pipe is given by,

$$h_{obs} = \frac{v^2}{2g}\left[\frac{A}{C_C(A-a)}\right]^2$$

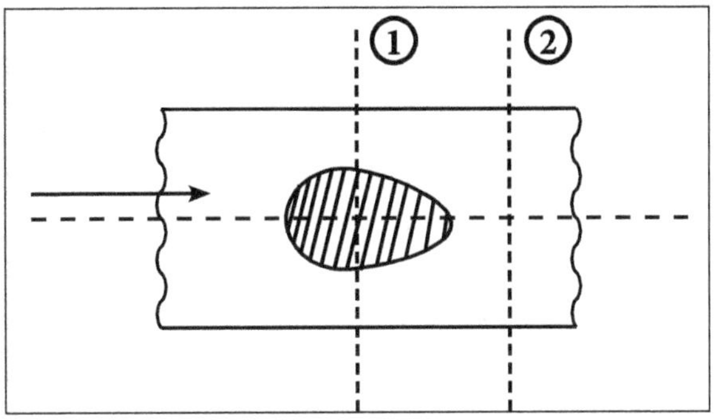

Loss of energy due to an obstruction in a pipe.

Where,

- $C_c \Rightarrow$ Coefficient of Contraction.
- $A \Rightarrow$ Area of C.S of the pipe.
- $a \Rightarrow$ Maximum area of C.S of the abstruction.

6. Loss of energy due to bend in pipe (h_b):

The loss of energy in bends is due to separation of flow from the boundary and the consequent formation of eddies resulting in the dissipation of energy.

Head loss due to bend:

$$h_b = \frac{KV^2}{2g}$$

Where, K⇒Coefficient of band.

7. Loss of energy in pipe fittings (h_v):

The loss of energy due to pipe fitting such as valves, couplings, etc. is expressed as:

$$h_v = \frac{KV^2}{2g}, \ K \Rightarrow \text{Coefficient}$$

Problems

1. Let us determine the loss of head when a pipe of diameter 200 mm is suddenly enlarged to a diameter of 400 mm. And also determine the rate of flow of water through the pipe is 250 liters/s.

Solution:

Given:

$D_1 = 200$ mm

$D_2 = 100$ mm

$Q = 250$ liters/s

To find: $h_e = ?$

Formula to be used:

$$D_1 = 200\,mm = 0.2\,m$$

$$A_1 = \frac{\pi}{4} \times D_1^2 = \frac{\pi}{4}(0.2)^2 = 0.03141\,m^2$$

$$D_2 = 400\,mm = 0.4\,m$$

$$A_2 = \frac{\pi}{4} \times (0.4)^2 = 0.12564\,m^2$$

$$\text{Discharge}, Q = 250\,\text{liters}/s = 0.25\,m^3/s,$$

$$V_1 = \frac{Q}{A_1} = \frac{0.25}{0.03141} = 7.96\,m/s$$

$$V_2 = \frac{Q}{A_2} = \frac{0.25}{0.12564} = 1.99\,m/s$$

Loss of head due to sudden enlargement:

$$h_e = \frac{(V_1 - V_2)^2}{2g} = \frac{(7.96 - 1.99)^2}{2 \times 9.81}$$

$$h_e = 1.816\,m\,of\,water$$

2. The rate of flow of water through a horizontal pipe is 0. 25 m^3/sec. The diameter of the pipe which is 20 cm is suddenly enlarged to 40 cm. The pressure intensity in the smaller pipe is 11.772 N/cm^2. Let us determine the loss of head due to sudden enlargement, pressure intensity in larger pipe and power loss due to enlargement.

Solution:

Given:

- Diameter of the smaller pipe, $D_1 = 200\,m = 0.2\,m$.
- Diameter of longer pipe, $D_2 = 40\,cm = 0.4\,m$.
- Rate of flow, of water, $Q = 0.25\,m^3/sec$.
- Intensity of pressure in the smaller pipe, $p_1 = 11.772\,N/cm^2$.

To find:

- Loss of head due to sudden enlargement, he.
- Intensity of pressure in the large pipe, p2.
- Power loss due to sudden enlargement Plost.

Formula to be used:

Now,

$$Velocity, v_1 = \frac{Q}{A_1} = \frac{0.25}{0.0314} = 7.961\,m/s$$

$$Velocity, v_2 = \frac{Q_2}{A_2} = \frac{0.25}{0.1256} = 1.990\,m/s$$

$$h_e = \frac{(v_1^2 - v_2)^2}{2g}$$

$$=\frac{(7.96-1.99)^2}{2\times 9.81}=1.816\,\text{m}$$

$$\frac{P_1}{w}+\frac{v_1^2}{2g}+z_1=\frac{P_2}{w}+\frac{v_2^2}{2g}+z_2+h_c$$

But, $z_1=z_2$ ...(because pipe is horizontal)

$$\therefore \frac{P_1}{w}+\frac{v_1^2}{2g}=\frac{P_2}{w}+\frac{v_2^2}{2g}+h_c$$

$$\text{or}\ \frac{P_2}{w}=\frac{P_1}{w}+\frac{v_1^2}{2g}-\frac{v_2^2}{2g}-h_c$$

$$P_2=9.81\left[\frac{11.772}{9.81}+\frac{7.91^2}{2\times 9.81}-\frac{1.990^2}{2\times 9.81}-1.816\right]$$

$$=9.81[1.2+3.19-0.202-1.816]$$

$$P_z=23.27\ \text{N/cm}^2$$

Intensity of pressure in the smaller pipe, $p_1=11.772\ \text{N/cm}^2$

Now,

$$\text{Velocity}, v_1=\frac{Q}{A_1}=\frac{0.25}{0.0314}=7.961\,\text{m/s}$$

$$\text{Velocity}, v_2=\frac{Q_2}{A_2}=\frac{0.25}{0.1256}=1.990\,\text{m/s}$$

1. Loss of head due to sudden enlargement, he

$$\frac{(v_1\quad v_2)}{2g}$$

$$=\frac{(7.96-1.99)^2}{2\times 9.81}=1.816\,\text{m}$$

2. Intensity of pressure in the large pipe, p2

Applying Bernoulli equation before and after sudden enlargement we get:

$$\frac{P_1}{w}+\frac{v_1^2}{2g}+z_1=\frac{P_2}{w}+\frac{v_2^2}{2g}+z_2+h_c$$

But, $z_1 = z_2$ ...(because pipe is horizontal)

$$\therefore \frac{P_1}{w}+\frac{v_1^2}{2g}=\frac{P_2}{w}+\frac{v_2^2}{2g}+h_c$$

$$\text{or}\ \frac{P_2}{w}=\frac{P_1}{w}+\frac{v_1^2}{2g}-\frac{v_2^2}{2g}-h_c$$

$$P_2=9.81\left[\frac{11.772}{9.81}+\frac{7.91^2}{2\times 9.81}-\frac{1.990^2}{2\times 9.81}-1.816\right]$$

$$=9.81\left[1.2+3.19-0.202-1.816\right]$$

$$P_z=23.27\ N/cm^2$$

3. Power loss due to sudden enlargement Plost

$$P_{lost}=\frac{WQh_c}{1000}kw$$

Where,

$$w=9.81\times 1000\ N/m^2$$

$$Q=0.25\ m^3/s$$

$$h_c = 1.816\,m$$

$$P_{lost}=\frac{(9.81\times 1000)\times 0.25\times 1.816}{1000}$$

$$=4.45\,kw$$

2.7.2 Flow Through Pipes in Series and Parallel

Equivalent Pipe

Equivalent pipe is defined as a pipe of uniform diameter having loss of head and discharge being equal to the loss of head and discharge of a compound pipe consisting of several pipes of different lengths and diameters.

The uniform diameter of the equivalent pipe is called equivalent size of the pipe. The length of equivalent pipe is equal to the sum of the lengths of the compound pipe consisting different pipes.

Let,

- L_1 = Length of pipe 1 and d1 = Diameter for pipe 1.
- L_2 = Length of pipe 2 and d2 =Diameter for pipe 2.
- L_3 = Length of pipe 3 and d3 = Diameter for pipe 3.
- L = Length of the equivalent pipe.
- d = Diameter of the equivalent pipe.

$$\frac{L_1}{d_1^5}+\frac{L_2}{d_2^5}+\frac{L_3}{d_3^5}=\frac{L}{d^5} \quad \text{(Dupuit's equation)}$$

In this equation $L=L_1+L_2+L_3$ from this diameter of equivalent pipe can be arrived.

A compound pipe consisting of several pipes of varying diameters and length may be replaced by a pipe of uniform diameter which is known as equivalent pipe.

For equivalent pipe, the loss of head and discharge are equal to the loss of head and discharge of compound pipe.

Let,

$$L_1, L_2, L_3 \Rightarrow \text{Length of Compound pipe}$$

$$D_1, D_2, D_3 \Rightarrow \text{Diameters of Compound}$$

$$H=\frac{4f}{2g}\left[\frac{L_1V_1^2}{D_1}+\frac{L_2V_2^2}{D_1}+\frac{L_3V_3^2}{D_1}\right] \qquad \ldots(1)$$

$$Q=A_1V_1=A_2V_2=A_3V_3$$

$$V_1 = \frac{Q}{A_1} = \frac{Q}{\frac{\pi}{4}D_1^2} = \frac{4Q}{\pi D_1^2}$$

Similarly,

$$V_2 = \frac{4Q}{\pi D_2^2}$$

$$V_3 = \frac{4Q}{\pi D_3^2}$$

Substituting V_1, V_2, V_3 in (1):

$$H = \frac{4f}{2g}\left[\frac{L_1}{D_1}\left(\frac{4Q}{\pi D_1^2}\right)^2 + \frac{L_2}{D_2}\left(\frac{4Q}{\pi D_2^2}\right)^2 + \frac{L_3}{D_3}\left(\frac{4Q}{\pi D_3^2}\right)\right]$$

$$H = \frac{4\times 16 + Q^2}{\pi^2 \times 2g}\left[\frac{L_1}{D_1^5} + \frac{L_2}{D_2^5} + \frac{L_3}{D_3^5}\right] \quad ...(2)$$

Head loss in the equivalent pipe:

$$H = \frac{4fLV^2}{2gnD}$$

Where,

$$V = \frac{4Q}{\pi D^2}$$

$$H = \frac{4fL}{2gD}\left(\frac{4Q}{\pi D^2}\right)^2$$

$$H = \frac{4\times 16fQ^2}{\pi^2 \times 2g}\left[\frac{L}{D^5}\right]$$

Substituting value of H in equation (2):

$$\frac{4\times 16fQ^2}{\pi^2 \times 2g}\left[\frac{2}{D^5}\right] = \frac{4\times 16fQ^2}{\pi^2 \times 2g}\left[\frac{L_1}{D_1^5} + \frac{L_2}{D_2^5} + \frac{L_3}{D_3^5}\right]$$

$$\frac{L}{D^5} = \frac{L_1}{D_1^5} + \frac{L_2}{D_2^5} + \frac{L_3}{D_3^5}$$

Where, $L = L_1 + L_2 + L_3$

The above equation is known as Dupuit's equation.

Pipes in Series

When pipes of different diameters are connected end to end to form a pipe line, they are said to be in series. The total loss of energy (or head) will be the sum of the losses in each pipe plus local losses at connections.

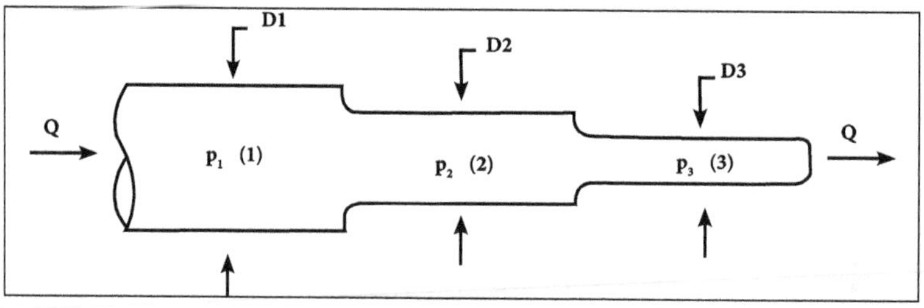

Pipes in series.

Pipes in Parallel

When two or more pipes in parallel connect two reservoirs as shown in Figure, for example, then the fluid may flow down any of the available pipes at different rates. But the head difference over each pipe will always be the same. The total volume flow rate will be the sum of the flow in each pipe. The analysis can be carried out by simply treating each pipe individually and summing flow rates at the end.

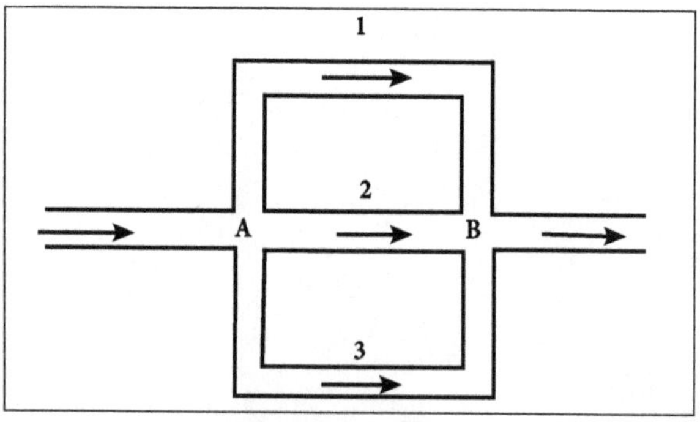

Pipes in parallel.

Problems

1. For a town water supply, a main pipe line of diameter 0.4 m is required. As pipes more than 0.35 m diameter are not readily available, two parallel pipes of same diameter are used for water supply. If the total discharge in the parallel pipes is same as in the single main pipe, let us determine the diameter of parallel pipe. Let us assume coefficient of discharge to be the same for all the pipes.

Solution:

Given: d = 0.4 m

To find:

The diameter of parallel pipe:

$$\text{Equating two losses } \frac{4fLV^2}{2g(0.4)} = \frac{4f \times L \times V^2}{2g \times d}$$

Formula to be used:

For a single pipe loss of head:

$$h = \frac{4fLV^2}{2g\,d} = \frac{4fLV^2}{2g(0.4)}$$

In case of parallel pipe, the diameters and lengths of the two pipes are the same. Hence discharge in each pipe will be half the discharge of single main pipe. As discharge in each parallel pipe is same, hence velocity will also be same.

- V = Velocity in each parallel pipe.
- d = Diameter of each parallel pipe.

Then, loss of head due to friction in parallel pipes:

$$h_* = \frac{4fLV_*^2}{2g \times d_*}$$

$$\text{Equating two losses } \frac{4fLV^2}{2g(0.4)} = \frac{4f \times L \times V_*^2}{2g \times d_*}$$

$$\Rightarrow \frac{V^2}{0.4} = \frac{V_*^2}{d_*}$$

$$\frac{0.4}{d_*} = \frac{4d_*^4}{0.0256}$$

$$d_*^5 = \frac{0.4 \times 0.0256}{4} = 0.00256$$

$$d_* = (0.00256)^{1/5} = 0.303\,\text{m}$$

$$d_* = 30.3\,\text{cm}$$

∴ Use two pipes of 30.3 cm diameter.

2. The two pipes of 15 cm and 30 cm diameters are laid in parallel to pass a total discharge of 100 liters per second. Each pipe is 250 m long. Let us determine discharge through each pipe. Now these pipes are connected in series to connect two tanks 500 m apart, to carry same total discharge. Let us also determine water level difference between the tanks. Neglect minor losses in both cases, f = 0.02 for both pipes.

Solution:

Given:

- $D_1 = 15$ cm
- $D_2 = 30$ cm
- $L_1 = L_2 = 250$ m
- Connect two tanks = 500 m
- $F = 0.02$

To find: Water level difference between the tanks.

Formula to be used:

$$Q_1 = \frac{\pi}{4} \times 0.15^2 \times V_1$$
$$= 0.0176\, V_1$$

$$H = \frac{4 f_1 L_1 V_1^2}{2 g d_1} + \frac{4 f_1 L_2 V_2^2}{2 g D_2}$$

$$= \frac{4 \times 0.02 \times 250 \times 1.415^2}{2 \times 9.81 \times 0.30} + \frac{4 \times 0.02 \times 250 \times 5.66^2}{2 \times 9.81 \times 0.15}$$

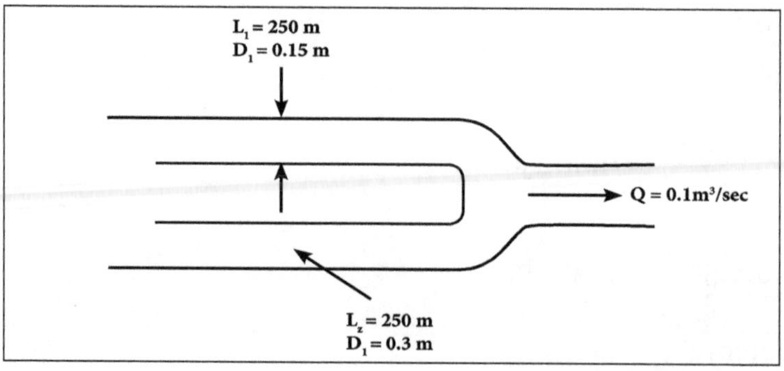

$$h_f = \frac{4f_1 L_1 V_1^2}{2gD_1} = \frac{4f_2 L_2 V_2^2}{2gD_2}$$

$$\frac{V_1^2}{l_1} = \frac{V_2^2}{l_2} \Rightarrow \frac{V_1^2}{0.15} = \frac{V_2^2}{0.3}$$

$$V_1 = \sqrt{0.5 V_2^2}$$

$$\frac{V_1^2}{0.15} = \frac{V_2^2}{0.3}$$

$$1.414 V_1 = V_2$$

Q_1 = Area of the Pipe (1) × Velocity in the Pipe (1)

$$= \frac{\pi}{4} \times 0.15^2 \times V_1$$

$$Q_1 = 0.0176 V_1$$

Q_2 = Area of the Pipe (2) × Velocity in the Pipe (2)

$$= \frac{\pi}{4} \times 0.3^2 \times V_2$$

$$= \frac{\pi}{4} \times 0.3^2 \times 1.414 V_1$$

$$Q_2 = 0.10 V_1$$

$$Q = Q_1 + Q_2$$

$$0.1 = 0.0176 V_1 + 0.10 V_1$$

$$V_1 = 0.85 \text{ m/sec} \Rightarrow V_2 = 1.414 V_1$$

$$= 1.414 \times 0.85$$

$$V_2 = 1.20 \text{ m/s}$$

Discharge in Pipe (1):

$$Q_1 = 0.0176 V_1$$

$$= 0.0176 \times 0.85$$

$$Q_1 \Rightarrow 0.015 \text{ m}^3/\text{sec}$$

Discharge in Pipe (2):

$$Q_2 = 0.10V_1 = 0.10 \times 0.85$$

$$Q_2 = 0.085\ m^3/sec$$

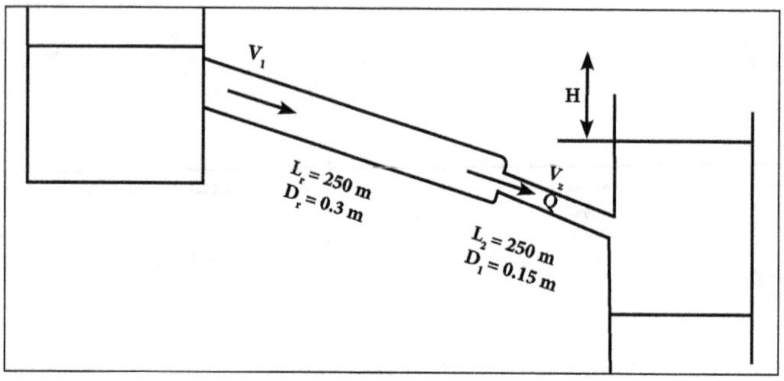

$$Q = 0.10 m^3/sec$$

$$Q = A_1 \times V_1$$

$$0.10 = \frac{\pi}{4} \times 0.3^2 \times V_1$$

$$V_1 = 1.415 m/sec$$

Similarly, $V_1 = \frac{0.10 \times 4}{\pi \times 0.32}$

$$Q = A_2V_2$$

$$1.10 = \frac{\pi}{4} \times 0.15^2 \times V_2$$

$$V_2 = \frac{0.10 \times 4}{\pi \times 0.15^2}$$

$$V_2 = 5.66 m/sec$$

Level difference:

$$H = \frac{4f_1L_1V_1^2}{2gd_1} + \frac{4f_1L_2V_2^2}{2gD_2}$$

$$= \frac{4 \times 0.02 \times 250 \times 1.415^2}{2 \times 9.81 \times 0.30} + \frac{4 \times 0.02 \times 250 \times 5.66^2}{2 \times 9.81 \times 0.15}$$

$$\Rightarrow 6.803 + 217.70$$

$$H = 224.50 \text{ m}$$

3. A main pipe dividing into two parallel pipes again forms one pipe. The length and diameter for the first parallel pipe are 2000 m and 1 m respectively, while the length and diameter of second parallel pipe are 2000 m and 0.8 m respectively. Let us determine the rate of flow in each parallel pipe, if total flow in the main is 3 m3/s. The coefficient of friction for each parallel pipe is same and equal to 0.005.

Solution:

Given:

- Length of Pipe (1) = 2000 m
- Diameter Pipe (1) = 1 m
- Length of Pipe (2) = 2000 m
- Diameter of Pipe (2) = 0.8 m

To find: The rate of flow in each parallel pipe.

Formula to be used:

$$Q = Q_1 + Q_2$$

Total flow $Q = 3 m^3/s$

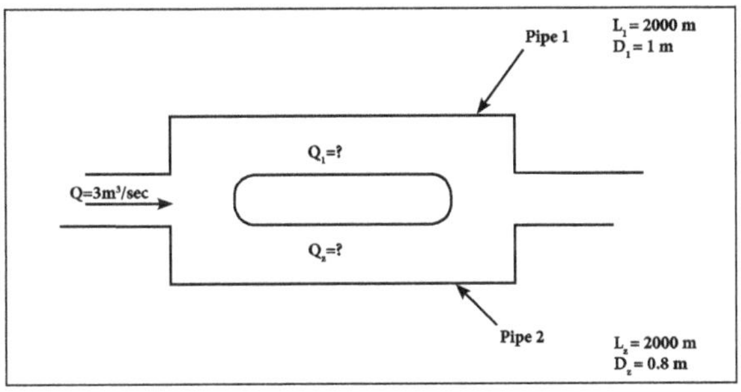

Co-efficient of friction, $f_1 = f_2 = 0.005$

$$h_f = \frac{4 f_1 L_1 V_1^2}{2gD_1} + \frac{4 f_2 L_2 V_2^2}{2gD_2}$$

$$0.8 V_1^2 = V_2^2$$

$$V_2 = \sqrt{0.8 V_1^2}$$

$$V_2 = 0.894 \text{ V}$$

$$Q_1 = \text{Area of Pipe } (1) \times \text{Velocity of Pipe } (1)$$

$$= \frac{\pi}{4} \times 1^2 \times V_1$$

$$Q_1 = 0.785 V_1$$

Similarly,

$$Q_2 = \frac{\pi}{4} \times 0.8^2 \times V_2$$

$$= \frac{\pi}{4} \times 0.8^2 \times 0.894 V_1$$

$$Q_2 = 0.45 V_1$$

But,

$$Q_1 + Q_2 = Q$$

$$785 V_1 + 0.45 V_1 = 3$$

$$V_1 = 2.43 \text{ m/s}$$

$$Q_1 = 0.785 V_1$$

$$= 0.785 \times 2.43$$

$$Q_1 = 1.907 \text{m}^3/\text{sec}$$

$$Q_2 = Q - Q_2 = 3 - 1.907$$

$$Q_2 = 1.093 \text{ m}^3/\text{s}$$

Result:

- Rate of flow in Pipe (1) = 1.907 m3/sec.
- Rate of flow in Pipe (2) = 1.093 m3/sec.

2.7.3 Hydraulic Gradient and Energy Gradient Line

The hydraulic grade line in open flow is the water surface and in pipe flow is the line which connects the elevations to which the water would rise in piezometer tube along the pipe. The energy gradient is at a distance equal to the velocity head above the hydraulic gradient. In both open and pipe flow the fall of the energy gradient for a given length of channel or pipe represents the loss of energy by friction.

When considered together, the hydraulic gradient and the energy gradient reflect not only the loss of energy by friction but also the conversions between potential and kinetic energy.

For uniform flow the hydraulic gradient and the energy gradient are parallel and the hydraulic gradient becomes an adequate basis for the determination of friction loss, since no conversion between kinetic and potential energy is involved.

For accelerated flow the hydraulic gradient is steeper than the energy gradient and in retarded flow the energy gradient is steeper than the hydraulic gradient. An adequate analysis of flow under these conditions cannot be made without consideration of both the energy gradient and the hydraulic gradient.

The Energy Line

The Energy Line is a line that represents the total head available to the fluid and can be expressed as:

$$EL = H = p/\gamma + v^2/2g + h = \text{Constant along a streamline}$$

Where, EL = Energy Line (m).

For a fluid flow without any losses due to friction (major losses) or components (minor losses) - The energy line would be at a constant level. In a practical world the energy line decreases along the flow due to losses.

A turbine in the flow reduces the energy line and a pump or fan in the line increases the energy line.

The Hydraulic Grade Line

The Hydraulic Grade Line is a line representing the total head available to the fluid - Minus the velocity head and can be expressed as:

$$HGL = p/\gamma + h$$

Where, HGL = Hydraulic Grade Line (m).

The hydraulic grade line lays one velocity head belowb the energy line.

3

Boundary Layer Theory and Dimensional Analysis

3.1 Introduction

A boundary layer is the layer of fluid in the immediate vicinity of a bounding surface where the effects of viscosity are significant.

Significance

In this region, the velocity gradient du/dy exists and hence the fluid exerts a shear stress on the wall in the direction of motion.

Table: Comparison between Stream line and Path line.

S. No.	Stream Line	Path Line
1.	It is an Imaginary line drawn through a flowing fluid in such a way that the tangent at end point on it indicates the velocity of that point.	A path line is a line that is traced by a single particle as it moves over a period of time. Path line shows the direction of velocity of the same particles at successive instant of time.
2.	The stream line does not cross.	Path line can intense itself at different times.

Equation for stream line in three dimensional flows:

$$\frac{dx}{u} = \frac{dy}{v} = \frac{dz}{w}$$

Drag

The component of the total force [FR] in the direction of motion is called drag. This component is denoted by [F_D]. Thus, drag is the force exerted by the fluid in the direction of motion.

Lift

The component of the total Force [FC] in the direction perpendicular to the direction of motion is known as lift. This is denoted by F_L. Thus, lift is the force exerted by the fluid in the direction perpendicular to the direction of motion.

3.1.1 Types of Boundary Layer Thickness

Laminar Boundary Layer Flow

The laminar boundary has a very smooth flow while the turbulent boundary layer contains swirls or "eddies", the laminar flow creates less skin friction drag than the turbulent flow, but is less stable. Boundary layer flow over a wing surface begins as a smooth laminar flow. As the flow continues back from the leading edge, the laminar boundary layer increases in thickness.

Turbulent Boundary Layer Flow

At some distance back from the leading edge, the smooth laminar flow breaks down and transitions to a turbulent flow. From a drag standpoint, it is advisable to have the transition from laminar to turbulent flow as far on the wing as possible or have a large amount of the wing surface within the laminar portion of the boundary layer. However, the low energy laminar flow tends to break down more suddenly than the turbulent layer.

The Boundary Layer Thickness

The boundary layer thickness, δ, is the distance across a boundary layer from the wall to a point where the flow velocity has essentially reached the 'free stream' velocity, mo. This distance is defined normal to the wall, and the point where the flow velocity is essentially that of the free stream is customarily defined as the point where:

$$u(y) = 0.99u_o$$

For laminar boundary layers over a flat plate, the Blasius solution gives:

$$\delta \approx 4.91\sqrt{\frac{vx}{u_o}}$$

$$\delta \approx \frac{4.91}{\sqrt{Re_x}}$$

For turbulent boundary layers over a flat plate, the boundary layer thickness is given by:

$$\delta \approx \frac{0.382x}{Re_x^{1/5}}$$

Where,

- $R_e = \rho u_o / x\mu$
- δ - is the overall thickness (or height) of the boundary layer.
- Re_x - is the Reynolds Number.
- ρ - is the density.
- u_o - is the free stream velocity.
- x - is the distance downstream from the start of the boundary layer.
- V - is the kinematic viscosity.
- μ - is the dynamic viscosity.

The velocity thickness can also be referred to as the Soole ratio, although the gradient of the thickness over distance would be adversely proportional to that of velocity thickness.

Displacement Thickness

A displacement thickness, δ* or δ1 is the distance by which a surface would have to be moved in the direction perpendicular to its normal vector away from the reference plane in an inviscid fluid stream of velocity uo to give the same flow rate as occurs between the surface and the reference plane in a real fluid.

In practical aerodynamics, the displacement thickness essentially modifies the shape of a body immersed in a fluid to allow an inviscid solution. It is commonly used in aerodynamics to overcome the difficulty inherent in the fact that the fluid velocity in the boundary layer approaches asymptotically to the free stream value as distance from the wall increases at any given location.

The definition of the displacement thickness for compressible flow is based on mass flow rate:

$$\delta^* = \int_0^{\infty} \left(1 - \frac{\rho(y)u(y)}{\rho_o u_o}\right) dy$$

The definition for incompressible flow can be based on volumetric flow rate, as the density is constant:

$$\delta^* = \int_0^{\infty} \left(1 - \frac{u(y)}{u_o}\right) dy$$

Where, ρ_o and u_o are the density and velocity in the 'free stream' outside the boundary layer and y is the coordinate normal to the wall.

For laminar boundary layers over a flat plate, the Blasius solution gives:

$$\delta^* \approx \frac{1.72x}{\sqrt{Re_x}}$$

For boundary layer calculations, the density and velocity at the edge of the boundary layer must be used, as there is no free stream. In the equations above, ρ_o and u_o are therefore replaced with ρ_e and u_e.

Problems

A smooth flat plate with a sharp leading edge is placed along a free stream of water flowing at 3 m/s. Let us calculate the distance from the leading edge and the boundary thickness where the transition from laminar to turbulent, flow may commence. Assume the density of water as 1000 kg/m³ and viscosity as 1 centipoise.

Solution:

Given:

- $U = 3\ m/s$
- $\rho = 1000\ kg/m^3$
- $\mu = 0.9$ centipoise

$$= \frac{0.9}{100} \text{poise}$$

$$= \frac{0.9}{1000} \frac{N-S}{m^2} = 9 \times 10^{-4}\ NS/m^2$$

To Find:

- Distance (X).
- Boundary layer thickness (δ).

Formula to be used:

$$R_e = \frac{\rho UX}{\mu}$$

$$\delta = \frac{5X}{\sqrt{R_{eX}}}$$

- Distance from the leading edge to the point where transition from laminar to turbulent occurs:

$$R_e = \frac{\rho UX}{\mu}$$

$$5\times10^5 = \frac{1000\times3\times X}{9\times10^{-4}}$$

$$\therefore X = 0.15\,\text{m}$$

- Thickness of Boundary at $X = 0.15\,\text{m}$:

$$\delta = \frac{5X}{\sqrt{R_{eX}}}$$

$$= \frac{5\times0.15}{\sqrt{5\times10^5}}$$

$$\delta = 1.06\times10^{-3}\,\text{m}$$

Result:

- Distance X = 0.15 m.
- Boundary layer thickness δ = 1.06 × 10−3 m.

3.1.2 Momentum and Energy Thickness

The nominal thickness of the boundary layer is defined as the thickness of zone extending from solid boundary to a point where velocity is 99% of the free stream velocity (U).

This is arbitrary, especially because transition from 0 velocity at boundary to the U outside the boundary takes place asymptotically. It is based on the fact that beyond this boundary, effect of viscous stresses can be neglected.

Momentum Thickness

Retardation of flow within boundary layer causes a reduction in the momentum flux too. So similar to displacement thickness, the momentum thickness (θ) is defined as the thickness of an imaginary layer in free stream flow which has momentum equal to the deficiency of momentum caused to actual mass flowing inside the boundary layer.

By equating the momentum flux rate for velocity defect to that for ideal fluid we get:

$$\rho U^2\theta = \int_0^\delta (\rho u dy)(U-u)$$

If density is constant:

$$\theta = \int_0^\delta \frac{u}{U}\left(1 - \frac{u}{U}\right) dy$$

θ would always be smaller than δ* and δ.

Graphical Representation

Energy Thickness

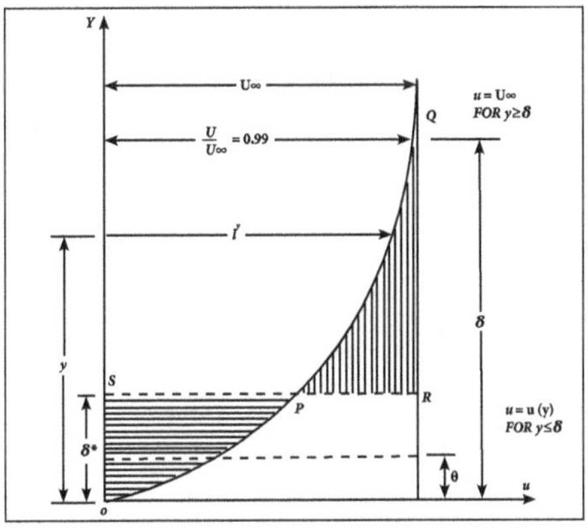

Similarly Energy thickness (δe) is defined as the thickness of an imaginary layer in free stream flow which has energy equal to the deficiency of energy caused to actual mass flowing inside the boundary layer.

Equation for Energy Thickness

By equating the energy transport rate for velocity defect to that for ideal fluid:

$$\frac{1}{2}\rho U^2 \delta_e = \frac{1}{2}\int_0^\delta (\rho u dy)\left(U^2 - u^2\right)$$

If density is constant, this simplifies to:

$$\delta_e = \int_0^\delta \frac{u}{U}\left(1 - \frac{u^2}{U^2}\right) dy$$

3.2 Separation of Boundary Layer and Control of Flow Separation

If the flow is reversed at the vicinity of the wall under certain conditions, the phenomenon is termed as separation of boundary layer.

Separation takes place due to excessive momentum loss near the wall in a boundary layer trying to move downstream against increasing pressure, i.e., $\frac{dp}{dx} > 0$, which is called adverse pressure gradient.

The figure below shows the flow past a circular cylinder, in an infinite medium.

Up to $\theta = 90^\circ$, the flow area is like a constricted passage and the flow behavior is like a nozzle.

Beyond $\theta = 90^\circ$ the flow area is diverged and the flow behavior is similar to a diffuser.

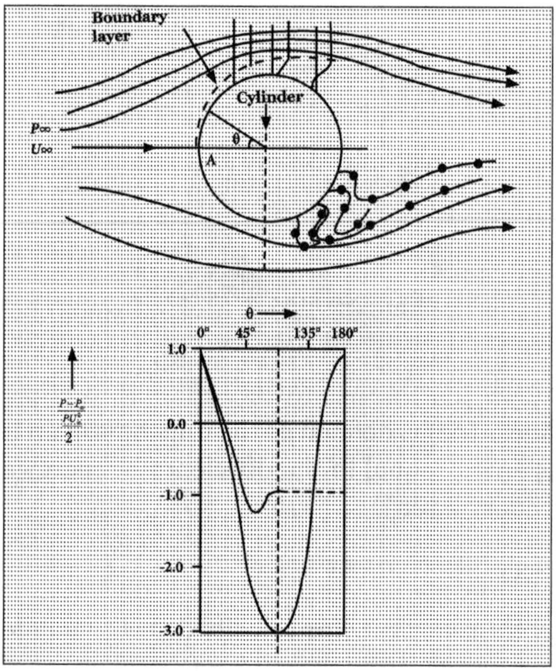

Flow separation and formation of wake behind a circular cylinder.

Here,

- P_∞: Pressure in the free stream.
- U_∞: Velocity in the free stream.
- ρ: Local pressure on the cylinder.

Consider the forces in the flow field. In the inviscid region:

Until $\theta = 90^\circ$, the pressure force and the force due to stream wise acceleration i.e. inertia forces are acting in the same direction (Pressure gradient is negative).

Beyond $\theta = 90^\circ$, the pressure gradient is positive or adverse. Due to the adverse pressure gradient the pressure force and the force due to acceleration will be opposite to each other.

So, long as no viscous effect is considered, the situation does not cause any sensation. In the viscid region (near the solid boundary).

Up to $\theta = 90^\circ$, the viscous force opposes the combined pressure force and the force due to acceleration. Fluid particles overcome this viscous resistance due to continuous conversion of pressure force into kinetic energy.

Beyond $\theta = 90^\circ$, within the viscous zone, the flow structure becomes different. It is seen that the force due to acceleration is opposed by both the viscous force and pressure force.

Depending upon the magnitude of adverse pressure gradient, somewhere around $\theta = 90^\circ$, the fluid particles in the boundary layer gets separated from the wall and driven in the upstream direction. The far field external stream pushes back these separated layers together with it and develops a broad pulsating wake behind the cylinder.

The point of separation may be defined as the limit between forward and reverse flow in the layer very close to the wall, i.e., at the point of separation:

$$\left(\frac{\partial u}{\partial y}\right)_{y=0} = 0 \qquad \text{... (1)}$$

This means that the shear stress at the wall, $\tau_w = 0$. But the adverse pressure continues to exist and at the downstream of this point the flow is in a reverse direction (back flow).

From the dimensional form of the momentum at the wall, where $u = v = 0$, we can write:

$$\left(\frac{\partial^2 u}{\partial y^2}\right)_{y=0} = \frac{1}{\mu}\frac{dp}{dx} \qquad \text{... (2)}$$

Consider the situation due to a favorable pressure gradient where $\frac{dp}{dx} < 0$ we have:

$\left(\partial^2 u / \partial y^2\right)_{wall} < 0$. (From Eq. (2))

As we proceed towards the free stream, the velocity u approaches U_∞ asymptotically, so $\partial u / \partial y$ decreases at a continuously lesser rate in y direction.

This means that $\partial^2 u / \partial y^2$ remains less than zero near the edge of the boundary layer.

The curvature of a velocity profile $\partial^2 u / \partial y^2$ is always negative. Consider the case of adverse pressure gradient, $\partial p / \partial x > 0$. At the boundary, the curvature of the profile must be positive (since $\partial p / \partial x > 0$).

Near the interface of boundary layer and free stream the previous argument regarding $\partial u / \partial y$ and $\partial^2 u / \partial y^2$ still holds good and the curvature is negative.

For an adverse pressure gradient, there must exist a point for which $\partial^2 u / \partial y^2 = 0$. This point is known as point of inflection of the velocity profile in the boundary layer. $\partial^2 u / \partial y^2 > 0$ at the wall since separation can only occur due to adverse pressure gradient. But we have already seen that at the edge of the boundary layer, $\partial^2 u / \partial y^2 < 0$. It is therefore; clear that if there is a point of separation, there must exist a point of inflection in the velocity profile.

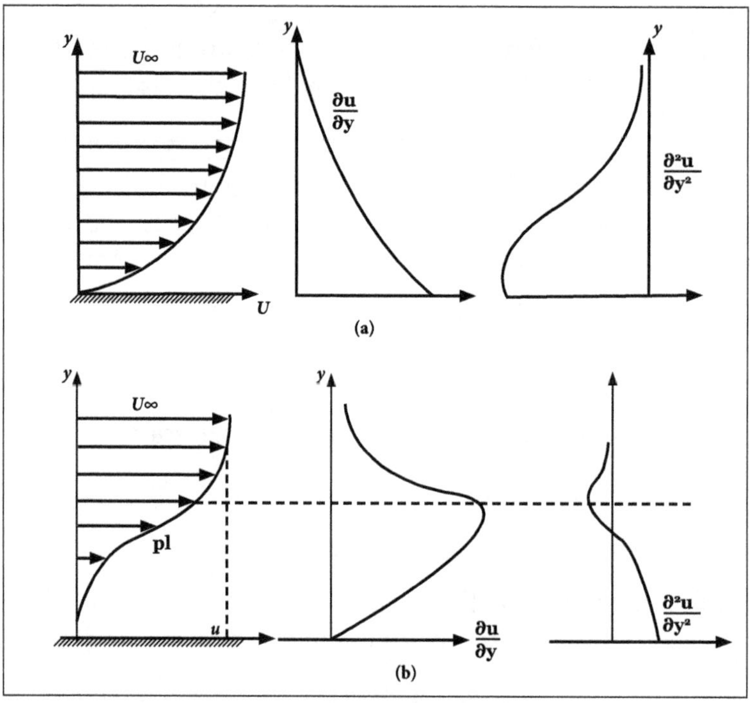

Velocity distribution within a boundary layer. (a) Favorable pressure gradient, $\frac{dp}{dx} < 0$ and (b) Adverse pressure gradient, $\frac{dp}{dx} > 0$.

Let us consider the flow past a circular cylinder and continue our discussion on the wake behind a cylinder. The pressure distribution which was shown by the firm line is obtained from the potential flow theory. However, somewhere near $\theta = 90^\circ$ (in experiments it has been observed to be at $\theta = 81^\circ$), the boundary layer detaches itself from the wall.

Meanwhile, pressure in the wake remains close to separation point pressure since the eddies cannot convert rotational kinetic energy into pressure head. The actual pressure distribution is shown by the dotted line, since the wake zone pressure is less than that of the forward stagnation point, the cylinder experiences a drag force which is basically attributed to the pressure difference.

The drag force, brought about by the pressure difference is known as form drag whereas the shear stress at the wall gives rise to skin friction drag. Generally, these two drag forces together are responsible for resultant drag on a body.

3.3 Stream Lined Body, Bluff Body and Basic Concepts of Velocity Profiles

Stream-Lined Body

A stream-lined body is defined as a body whose surface coincides with the stream-lines, when it is placed in a flow. In that case the flow separation will take place only at the trailing edge (or rearmost part). Though the boundary layer will start at the leading edge, it will become turbulent from laminar. Still it does not separate up to the rear-most pan of the body in the case of stream-lined body.

Thus behind a stream-lined body, wake formation zone will be small and therefore the pressure drag will also be small. The total drag on the stream-lined body will be due to friction (shear) only. A body may be stream-lined:

- When placed in a particular position in the flow but may not be so when placed in another position.
- At low velocities but may not be so at higher velocities.

Bluff Body

A bluff body is defined as a body whose surface does not coincide with the stream-lines, when placed in a flow. Then the flow is separated from the surface of the body much ahead of its trailing edge with a very large wake formation zone. The drag due to pressure will be large compared to the friction drag on the body. Thus the bodies in which the pressure drag is very large compared to friction drag are called bluff bodies.

These bodies play a very important role in designing the shape profile of cars, aerospace objects, etc. where care must be taken for aerodynamics profile.

- Flows with large wakes (comparable to the typical dimension of the body) and significant changes when compared to the ideal fluid model (large δ^*) originating a high pressure (form) drag force.

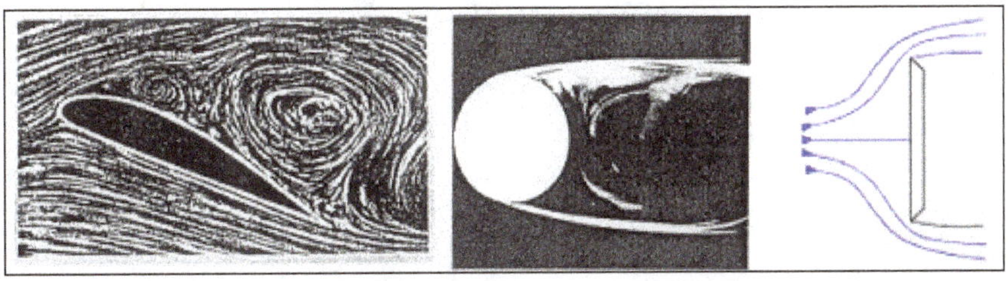

- Large wake is related to the body shape (cylinder, sphere) or with the orientation of the incoming flow (flat plate, foil).

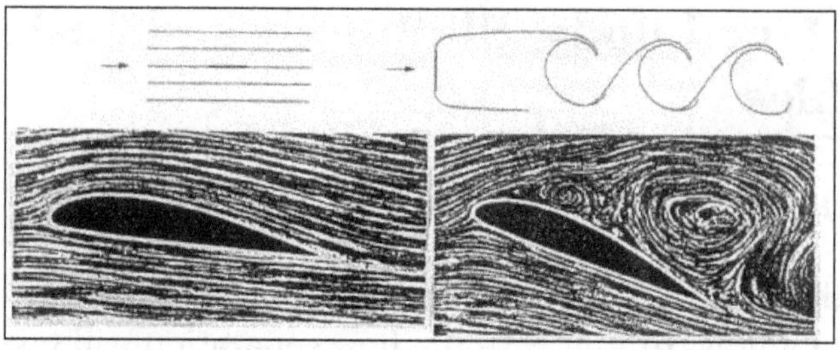

- Near wake have small velocities and an approximately constant pressure smaller than the undisturbed pressure.
- Far wake exhibits vortices of symmetric strength aligned along two parallel lines and shifted half wavelength "von Karman Street".
- Vortex shedding leads to an unsteady flow and problems of induced vibrations.

3.4 Dimensional Analysis

The dimensionless analysis is a mathematical technique used to predict physical parameters that influence the flow in fluid mechanics, heat transfer in thermodynamics and so forth. The analysis involves the fundamental units of dimensions MLT: Mass, Length and Time. It is helpful in experimental work because it provides a guide to factors that significantly affect the studied phenomena.

Fundamental Units

Fundamental units are the three fixed dimensions, i.e. Length L, Mass M and Time T which is of importance in fluid mechanics.

Derived Units

Secondary or Derived quantities are those quantities which possess more than one fundamental dimension.

Example: Density $\left(\frac{M}{L^3}\right)$

Dimensionally Homogeneous Equation

A physical equation is the relationship between two or more physical quantities. Any correct equation expressing a physical relationship between quantities must be dimensionally homogeneous (According to Fourier's principle of dimensional homogeneity) and numerically equivalent.

Dimensional homogeneity states that every term in equation, when reduced to fundamental dimensions must contain identical powers of each dimension. A dimensionally homogeneous equation is applicable to all system.

Example: Let us consider the following equation:

$$P = \omega h$$

Dimensions of L.H.S. = $ML^{-1}T^{-2}$.

Dimensions of R.H.S. = $ML^{-2}T^{-2} \times ML^{-1}T^{-2}$.

Dimensions of L.H.S. = Dimensions of R.H.S.

∴ Equation $P = \omega h$ is dimensionally homogeneous. So it can be used in any system of units.

Velocity Potential and Stream Function

The velocity potential is defined as a scalar function of space and time such that its negative derivative with respect to any direction gives the fluid velocity in that direction, it is denoted by φ (phi).

Thus mathematically the velocity potential is defined as:

$$\varphi = f(x, y, z, t) \quad \text{... For unsteady flow}$$

$$\varphi = f(x, y, z) \quad \text{... For steady flow}$$

Stream Function

The stream function is defined as a scalar function of space and time, such that its partial derivative with respect to any direction gives the velocity component at sight angle to this direction. It is denoted by ψ (psi).

In case of two-dimensional flow, the stream function may be defined. Mathematically are:

$$\psi = f(x, y, t) \quad \text{... For unsteady flow}$$

$$\psi = f(x, y) \quad \text{... For steady flow}$$

3.4.1 Methods of Dimensional Analysis

The Buckingham's Π-Theorem

"If there are n variables (dependent and independent variables) in a dimensionally homogeneous equation and if these variables contain m fundamental dimensions (such as M, L, T, etc.) then the variables are arranged into (n-m) dimensionless terms. These dimensionless terms are called T-terms".

Mathematically, if any variable X1 depends on independent variables X_2, X_3, X_4.....X_n in the functional equation and may be written as:

$$X_1 = f(X_2, X_3, X_4, ..., X_n)$$

The above equation can also be written as:

$$f_1(X_1, X_2, X_3, ..., X_n) = 0$$

Selection of Repeating Variables

The following points should be kept in view while selecting "m" repeating variables:

- "m" repeating variables must contain jointly all the fundamental dimensions involved in the phenomenon, usually the fundamental dimensions are M, L and T. However, if only two dimensions are involved.
- The repeating variables must not form the non-dimensional parameters among themselves.
- As far as possible, the dependent variable should not be selected as repeating variable.
- No two repeating variables should have the same dimensions.
- The repeating variables should be chosen in such a way that one variable contains geometric property (e.g., Length (L), diameter (d), height (H), etc.) and second variable contains flow property (e.g., velocity (v), acceleration (a), etc.) and third variable contains fluid property (e.g., mass density (ρ), weight density (w), dynamic viscosity (μ), etc.).

Table: Comparison between Buckingham's Π-Theorem and Rayleigh's Method.

S. No.	Rayleigh's Method	Buckingham's n-Theorem
1.	This method is used for determining the expression for a variable which depends upon maximum 3 or 4 variables only.	This method is used for determining the expression for n number of variables.
2.	It is more difficult if the variables are more than number of fundamental dimensions.	No such difficulties occur.

Problems

1. Let us check whether the following equation is dimensionally homogeneous.

Solution:

Given:

Equation Q $Cd \cdot a\sqrt{2gh}$

Dimensions of L.H.S. $\Rightarrow Q = L^3T^{-1}$

Dimensions of R.H.S. $\Rightarrow$ Cd a

$$= L^2 \times \sqrt{\frac{L}{T^2} \times L}$$

$$= L^2 \times \sqrt{\frac{L^2}{T^2}} = L^3 T^{-1}$$

2. A drag force (F) on a partially submerged body depends on the relative velocity (V) between the body and the fluid, characteristic linear dimension (1), height of surface roughness (k), fluid density (p), the viscosity (p) and the acceleration due to gravity (g). Let us obtain an expression for the drag force, using the method of dimensional analysis.

Solution:

Given: (i) Force, (ii) Velocity, (iii) Length, (iv) K, (v) ρ, (vi) μ and (vii) g.

To find: Equation for drag force.

Formula to be used:

$$F = MLT^{-2}, V = LT^{-1}, \; L = L, \; K = L, \; \rho = ML^{-3}, \; \mu = ML^{-1}T^{-1} \text{ and } g = LT^{-2}$$

By Mathematical Equation:

$$f(F, v, l, K, \rho, \mu, g)$$

$$n = 7$$

$$F = MLT^{-2}, V = LT^{-1}, \; L = L, \; K = L, \; \rho = ML^{-3}, \; \mu = ML^{-1}T^{-1} \text{ and } g = LT^{-2}$$

Using Buckingham π theorem:

$$\pi = n - m$$

$$\pi = 7 - 3$$

$$\pi = 4$$

$$f(\pi_1, \pi_2, \pi_3, \pi_4) = 0$$

π_1 term:

$$\pi_1 = \left(\overset{a_1}{\rho}, \overset{b_1}{V}, \overset{c_1}{l}, g\right)$$

$$\pi_2 = \left(\overset{a_2}{\rho}, \overset{b_2}{V}, \overset{c_2}{l}, K\right)$$

$$\pi_3 = \left(\overset{a_3}{\rho}, \overset{b_3}{V}, \overset{c_3}{l}, \mu\right)$$

$$\pi_4 = \left(\overset{a_4}{\rho}, \overset{b_4}{V}, \overset{c_4}{l}, F\right)$$

π_1 term:

$$M^oL^oT^o = \left(ML^{-3}\right)^{a_1}\left(LT^{-1}\right)^{b_1} L^{c_1}LT^{-2}$$

$$M \Rightarrow o = a_1$$

$$L \Rightarrow o = -3a_1 + b_1 + c_1 + 1 \Rightarrow c_1 = 1$$

$$T \Rightarrow o = -b_1 - 2 \Rightarrow b_1 = -2$$

$$\pi_1 = \frac{gl}{V^2}$$

π_2 term:

$$M^oL^oT^o = \left(ML^{-3}\right)^{a_2}\left(LT^{-1}\right)^{b_2} L^{c_2}L$$

$$M \Rightarrow o = a_1$$

$$L \Rightarrow o = -3a_1 + b_1 + c_1 + 1 \Rightarrow c_1 = 1$$

$$T \Rightarrow o = -b_1$$

$$\pi_1 = \frac{K}{l}$$

π_3 term:

$$M^oL^oT^o = \left(ML^{-3}\right)^{a_3}\left(LT^{-1}\right)^{b_3} L^{c_3}ML^{-1}T^{-1}$$

$$M \Rightarrow o = a_3 + 1 \Rightarrow a_3 = -1$$

$$L \Rightarrow o = -3a_3 + b_3 + c_3 - 1 \Rightarrow c_3 = -1$$

$$T \Rightarrow o = -b_3 - 1 \Rightarrow b_3 = -1$$

$$\pi_3 = \frac{\mu}{\rho V l}$$

π_4 term :

$$M^oL^oT^o = \left(ML^{-3}\right)^{a_3}\left(LT^{-1}\right)^{b_4} L^{c_4} MLT^{-2}$$

$$M \Rightarrow 0 = a_4 + 1 \Rightarrow a_4 = -1$$

$$L \Rightarrow 0 = -3a_4 + b_4 + c_4 + 1 \Rightarrow c_4 = -2$$

$$T \Rightarrow 0 = -b_4 - 1 \Rightarrow b_4 = -2$$

$$\pi_4 = \frac{F}{\rho V^2 l^2}$$

$$f\left(\pi_1, \pi_2, \pi_3, \pi_4\right) = 0$$

$$f\left(\frac{gl}{V^2}, \frac{K}{l}, \frac{\mu}{\rho Vl}, \frac{F}{l^2 V^2 l^2}\right) = 0$$

$$\frac{F}{\rho V^2 l^2} = \frac{V^2}{gl} \times \frac{l}{k} \times \frac{\rho Vl}{\mu}$$

$$F = \rho V^2 l^2 \phi\left(\frac{\rho V^3 l}{\mu Kg}\right)$$

3.5 Similitude and Modeling

Similitude is a concept applicable to the testing of engineering models. A model is said to have similitude with the real application if the two share geometric similarity, kinematic similarity and dynamic similarity. Similarity and similitude are interchangeable in this context.

The term dynamic similitude is often used as a catch all because it implies that geometric and kinematic similitude have already been met.

Similitude's main application is in hydraulic and aerospace engineering to test fluid flow conditions with scaled models.

Engineering models are used to study complex fluid dynamics problems where calculations and computer simulations aren't reliable. Models are usually smaller than the final design but not always. Scale models allow testing of a design prior to building and in many cases are a critical step in the development process.

The construction of a scale model however must be accompanied by an analysis to determine what conditions it is tested under. While the geometry may be simply scaled, other parameters such as pressure, temperature or the velocity and type of fluid may need to be altered. Similitude is achieved when testing conditions are created such that the test results are applicable to the real design.

Types of Similitude

The following three types of similarities must be maintained between the model and prototype to facilitate useful study of the model investigation.

- Geometric similarity.
- Kinematic similarity.
- Dynamic similarity.

Geometric Similarity:

This similarity between the model and the prototype is said to exist if the linear dimensions between the two are equal. If Lm. $L_p, B_m, B_p, D_m, D_p, \forall_m, \forall_p, A_m, A_p, d_m, d_p$ are the length, breadth, diameter, volume, area, depth of model (with subscript in) and prototype (with subscript p). respectively.

Then,

$$\frac{L_m}{L_p} = \frac{D_m}{D_p} = \frac{B_m}{B_p} = Lr \text{ (i.e. the length ratio)}$$

$$\frac{A_m}{A_p} = \frac{L_m \times B_m}{L_p \times B_m} = Lr \times Lr = Lr^2 \text{ (the area ratio)}$$

$$\frac{\forall_m}{\forall_p} = \frac{L_m \times B_m \times D_m}{L_p \times B_p \times D_p} = Lr \cdot Lr \cdot Lr = Lr^3 \text{ (volume ratio)}$$

If the above ratios of the model and the prototype are maintained, it is said that geometric similarity exists.

Kinematic Similarity:

If the ratio of kinematic parameters like velocity, acceleration, time and discharge at the corresponding points in the model and the prototype are the same, it is said that there is kinematic similarity between the model and the prototype. Since velocity and acceleration are vector quantities, the directions of velocity and acceleration at corresponding points in the model and prototype must be parallel.

$$T_r = \frac{T_m}{T_p}, \text{ time scale ratio}$$

$$\frac{V_m}{V_p} = V_r, \text{ velocity scale ratio}$$

$$= (L_m/T_m)(L_p/T_p)$$

$$\frac{a_m}{a_p} = a_r, \text{ acceleration scale ratio}$$

$$= \frac{\left(\frac{L_m}{T_{m^2}}\right)}{\left(\frac{L_p}{T_{p^2}}\right)} = \left(\frac{L_m}{L_p}\right)\left(\frac{T_p^2}{T_{n^2}}\right) = L_r/T_r$$

$$\frac{Q_m}{Q_p} = Q_r = \frac{\left(\frac{L_m^3}{T_m}\right)}{\left(\frac{L_p^3}{T_p}\right)} = \left(\frac{L_m}{L_p}\right)^3 \cdot \frac{T_p}{T_n} = L_r^3 / T_r$$

Kinematic similarity is attained if the flow net formed by streamlines and the equipotential lines for the model and the prototype are geometrically similar. Changing in the scale and flow net of the model and the prototype can be superimposed.

Dynamic Similarity:

If both geometrical and kinematic similarities exist between the model and the prototype, then dynamic similarity for the model and the prototype is attained. This means that the ratio of all forces acting on homologous points in the model and the prototype are equal. The forces acting on the fluid flow system are:

- Inertia force F_i.
- Viscous force F_v.
- Gravity force F_g.
- Surface tension force F_s.
- Elastic force F_e.
- Pressure force F_p.

Let us consider that the 'x' component of velocity is $u = x^2 + z^2 + 5$ and the 'y' component is $v = y^2 + z^2$. And let us find the simplest component of velocity that satisfies continuity.

The continuity equation for in-compressible fluid is given by:

$$\frac{\partial u}{\partial x} + \frac{\partial v}{\partial y} + \frac{\partial w}{\partial z} = 0$$

$$u = x^2 + z^2 + 5$$

$$v = y^2 + z^2$$

$$\frac{\partial u}{\partial x} = 2x$$

$$\frac{\partial v}{\partial y} = 2y$$

Substituting the values of $\frac{\partial u}{\partial x}$ and $\frac{\partial v}{\partial y}$ equation in continuity. We get:

$$2x + 2y + \frac{\partial w}{\partial z} = 0$$

$$\frac{\partial w}{\partial z} = -2x - 2y$$

Integration on both sides:

$$\int dw = \int(-2x - 2y)dz$$

$$w = -2xz - 2yz + \text{constant of integration}$$

Where, constant of integration cannot be a function of z. But it can be a function of x and y that is f(x, y).

$$w = (-2xz - 2yz) + f(x, y)$$

3.6 Dimensionless Numbers

In dimensional analysis, a dimensionless quantity is a quantity to which no physical dimension is applicable. It is thus a bare number and is therefore known as a quantity of dimension one.

Reynolds Number

It is defined as the ratio of inertia force of a flowing fluid to the viscous force of the fluid. The expression for Reynolds number is obtained as:

Inertia force (F_i) = Mass × Acceleration of flowing fluid

$$= \rho \times \text{Volume} \times \frac{\text{Velocity}}{\text{Time}}$$

$$= \rho \times \frac{\text{Volume}}{\text{Time}} \times \text{Velocity}$$

$$= \rho \times AV \times V \quad [\because \text{Volume per sec.} = \text{Area Velocity}]$$

$$= AV$$

$$= \rho AV^2$$

Viscous Force (F_v) = Shear Stress × Area

$$= \tau \times A$$

$$= \left(\mu \frac{du}{dy}\right) \times A$$

$$= \mu \cdot \frac{V}{L} \times A$$

By definition Reynolds number is:

$$R_e = \frac{F_i}{F_v}$$

$$= \frac{\rho A V^2}{\mu \times \frac{V}{L} \times A} = \frac{\rho V L}{\mu}$$

$$= \frac{VXL}{(ì/ñ)} = \frac{VL}{\gamma} \quad (\because \gamma = ì/ñ)$$

In case of pipe flow, the linear dimension L is taken as diameter d. Hence Reynolds number for pipe flow:

$$R_e = \frac{V_d}{\gamma} \text{ (or) } \frac{\rho v d}{\mu}$$

Froude's Number

The Froude's number is defined as the square root of the ratio of inertia force of a flowing fluid to the gravity force. Mathematically it is expressed as:

$$F_e = \sqrt{\frac{F_i}{F_g}}$$

$$\text{Inertia force}(F_i) = \rho A V^2$$

$F_g \Rightarrow$ Force due to gravity

$\therefore F_g$ = Mass × Acceleration due to gravity

$$= \rho \times \text{Volume} \times g = \rho \times L^3 \times g$$

$$= \rho \times L^2 \times L \times g = \rho \times A \times L \times g$$

$$\therefore F_e = \sqrt{\frac{F_i}{F_g}}$$

$$= \sqrt{\frac{\rho AV^2}{\rho AL_g}}$$

$$= \sqrt{\frac{V^2}{L_g}} = -\frac{V}{\sqrt{L_g}}$$

Application

It is used in open hydraulic structure such as spillways, weirs, open channel flow, slices etc. in which force due to gravity are predominant.

Mach's Number

Mach's number is defined as the square root of the ratio of the inertia force of a flowing fluid to the elastic force. Mathematically it is defined as:

$$M = \sqrt{\frac{\text{Inertia force}}{\text{Elastic force}}} = \sqrt{\frac{F_i}{F_e}}$$

Where,

$$F_i = \rho AV^2$$

$$F_e = \text{Elastic force}$$

$$= \text{Elastic stress} \times \text{Area}$$

$$= K \times A$$

$$= KL^2$$

$$M = \sqrt{\frac{\rho AV^2}{KL^2}} = \sqrt{\frac{\rho L^2 V^2}{KL^2}}$$

$$= \sqrt{\frac{V^2}{k/\rho}} = \sqrt{\frac{V}{k/\rho}}$$

But,

$$\sqrt{k/\rho} = C$$

$$M = \frac{V}{C}$$

Application

In compressible flow, it is used at high speed such as high velocity flow in pipes, water hammer, motion of high speed objects like the aero plane, projectile and missiles, where the elastic force (compressibility effect) are predominant.

Weber's Number

It is defined as the square root of the ratio of inertia force of a flowing fluid to the surface tension force. Mathematically it is expressed as:

$$W_e = \sqrt{\frac{F_i}{F_S}}$$

Where,

$$F_i = \text{Inertia force} = \rho A V^2$$

$$F_S = \text{Surface tension force}$$

$$= \text{Surface tension per unit length} \times \text{Length}$$

$$= \sigma \times L$$

$$W_e = \sqrt{\frac{\rho A V^2}{\sigma \times L}} = \sqrt{\frac{\rho \times L^2 \times V^2}{\sigma \times L}}$$

$$= \sqrt{\frac{\rho L \times V^2}{\sigma}}$$

$$= \sqrt{\frac{V^2}{\sigma / \rho L}}$$

$$= \frac{V}{\sqrt{\sigma / \rho L}}$$

Applications

Capillary movement of water in soil, flow of blood in veins and arteries where the surface tension effects are predominant.

Problems

1. Let us express the efficiency in terms of dimensionless parameters using density, viscosity, angular velocity, diameter of the rotor and discharge using Buckingham Π - theorem.

Solution:

η is a function of ρ, μ, ω, D and Q

$$\eta = f\ (\rho,\ \mu,\ \omega,\ D,\ Q)$$

$$f_1(\eta,\ \rho,\ \mu,\ \omega,\ D,\ Q) \quad ...(1)$$

Hence total no. of variables, n + 6

No. of fundamental dimensions, m = 3

Dimensions of each variable are:

$$\eta = \text{Dimension less}$$

$$\rho = ML^{-3}$$

$$\omega = T^{-1}$$

$$D = L$$

$$Q = L^3 T^{-1}$$

$$\therefore m = 3$$

No. of Π terms $= n - m = 6 - 3 = 3$

$\therefore$ (1) becomes:

$$f_1(\pi_1, \pi_2, \pi_3) = 0$$

Each π term contains m + 1 variables and m number of repeating variables.

Choosing D, ω and ρ and repeating variables and then we have:

$$\pi_1 = D^{a_1} \cdot \omega^{b_1} \cdot \rho^{c_1} \cdot \eta$$
$$\pi_2 = D^{a_2} \cdot \omega^{b_2} \cdot \rho^{c_2} \cdot \mu$$
$$\pi_3 = D^{a_3} \cdot \omega^{b_3} \cdot \rho^{c_3} \cdot Q$$

First π term (π_1):

$$\pi_1 = D^{a_1} \cdot \omega^{b_1} \cdot \rho^{c_1} \cdot \eta$$
$$M^0L^0T^0 = L^{a_1} \cdot (T^{-1})^{b_1} \cdot (ML^{-3})^{c_1} \cdot M^0L^0T^0$$

Equating powers on both sides:

$$\text{Power of M}, 0 = c_1 + 0 \Rightarrow c_1 = 0$$
$$\text{Power of L}_1\ 0 = a_1 + 0_1 \Rightarrow a_1 = 0$$
$$\text{Power of T}_1\ 0 = -b_1 + 0 \Rightarrow b_1 = 0$$

Substituting the values of a1, b1 and c1 in π and then we get:

$$\pi_1 = D^{o} \cdot \omega^{o} \cdot \rho^{o} \cdot \eta = \eta$$

Second π Term

Substituting dimensions on both sides:

$$\pi_2 = D^{a_2} \cdot \omega^{b_2} \cdot \rho^{c_2} \cdot \mu$$

Equating powers on both sides:

$$\text{Power of M}, 0 = c_2 + 1 \Rightarrow c_2 = -1$$
$$\text{Power of L}, 0 = a_2 - 3c_2 - 1$$
$$\therefore a_2 = 3c_2 + 1$$
$$= -3 + 1 = -2$$
$$\text{Power of T}, 0 = -b_2 - 1 \Rightarrow b_2 = -1$$

Substituting the values of a_2, b_2 and c_2 in T_1:

$$\pi_2 = D^{-2} \cdot \omega^{-1} \cdot \rho^{-1}\, \mu = \frac{\mu}{D^2 \omega \rho}$$

$$M^{o} L^{o} T^{o} = L^{a_2} \cdot \left(T^{-1}\right)^{b_2} \cdot \left(ML^{-3}\right)^{c_2} \left(ML^{-1}T^{-1}\right)$$

Third π term:

$$\pi_3 = D^{a_3} \cdot \omega^{b_3} \cdot \rho^{c_3} \cdot Q$$

Substituting the dimensions on both sides:

$$M^{o} L^{o} T^{o} = L^{a_3} \cdot \left(T^{-1}\right)^{b_3} \cdot \left(ML^{-3}\right)^{c_3} L^3 T^{-1}$$

Equating the powers of M, L and T on both sides:

$$\text{Power of M}, 0 = c_3, \Rightarrow c_3 = 0$$
$$\text{Power of L}, 0 = a_3 - 3c_3 + 3$$
$$\Rightarrow a_3 = 3c_3 - 3 = -3$$

$$\text{Power of T}, 0 = -b_3 - 1$$
$$b_3 = -1$$

Substituting the values of a_3, b_3 and c_3 in π_3.

$$\pi_3 = D^{-3} \cdot D^{-3} \cdot \rho^0 - Q = \frac{Q}{D^3\omega}$$

Substituting the values $\pi 1$, π_2 and π_3 is equation (ii):

$$f_1\left(\eta \cdot \frac{\mu}{D^2\omega\rho} \cdot \frac{Q}{D^3\omega}\right) = 0$$

$$\eta = \phi\left[\frac{\mu}{D^2\omega\rho} \cdot \frac{Q}{D^3\omega}\right]$$

- Application of Dimensionless Parameter
- Dimensionless groups can be used to describe complicated processes which are influenced by a number of variables.
- The whole process can be analyzed on a sound basis by means of a few dimensionless parameters.

Problems

1. A resisting force (R) of a supersonic flight being considered as dependent upon the length of the air craft 'l' velocity 'v', air viscosity 'μ' air density 'ρ' and bulk modulus of air is 'k'. Let us express the functional relationship between these variables and the resisting force.

Solution:

Given:

The resisting force K depends upon:

- Length, l
- Velocity, V
- Viscosity, μ
- Bulk modulus, K
- Density, ρ

To find: The functional relationship between these variables and the resisting force.

Formula to be used:

$$MLT^{-2} = ALa\left(LT^{-1}\right)b \cdot \left(ML^{-1}T^{-1}\right)c \cdot \left(ML^{-3}\right)d \cdot \left(ML^{-1}\ T^{-2}\right)e$$

$$K\,Ala \cdot Vb \cdot \mu c \cdot \rho d \cdot Ke \qquad \text{... (1)}$$

Where, A is the non-dimensional constant.

Substituting the dimensions on both sides of the equation (1):

$$MLT^{-2} = ALa\left(LT^{-1}\right)b \cdot \left(ML^{-1}T^{-1}\right)c \cdot \left(ML^{-3}\right)d \cdot \left(ML^{-1}\ T^{-2}\right)e$$

Equating the Powers of M, L, T on both sides:

$$\text{Power of } M, 1 = a + b - c - 3d - e$$
$$\text{Power of } L, 1 = c + d + e$$
$$\text{Power of } T, -2 = -\,b - c - 2e$$

There are five unknowns but equations are only three. Express the three unknowns in terms of two unknowns (μ and K).

∴ Express the value of a, b and d in terms of c and e.

Solving,

$$d = 1 - c - e$$

$$b = 2 - c - 2e$$

$$a = 1 - b + c + 3d + e$$

$$= 1 - (2 - c - 2e) + c + 3(1 - c - e) + e$$

$$= 1 - 2 + c + 2e + c + 3 - 3c - 3e + e$$

$$= 2 - c$$

Substituting these values in (1) we get:

$$R = Al^2\ -c \cdot V\ ^2 - c - 2e \cdot \mu c \cdot \rho 1 - c - e \cdot Ke$$
$$= Al^2 \cdot V\ ^2 \cdot \rho\left(1 - c \cdot V - c \cdot \mu c \cdot \rho - c\right) \cdot \left(V - 2e \cdot \rho - e \cdot Ke\right)$$
$$= Al^2\ V^2 \rho\left(\frac{\mu}{\rho VL}\right)^{\rho} \cdot \left(\frac{K}{\rho V^2}\right)^{\rho}$$
$$R = A\rho l^2\ V^2 \phi\left[\left(\frac{M}{\rho VL}\right)^{\rho} \cdot \left(\frac{K}{\rho V^2}\right)^{\rho}\right]$$

Turbo Machinery

4

4.1 Basics of Turbo Machinery

4.1.1 Jet forces on Stationary Plates

Jet force on a flat plate:

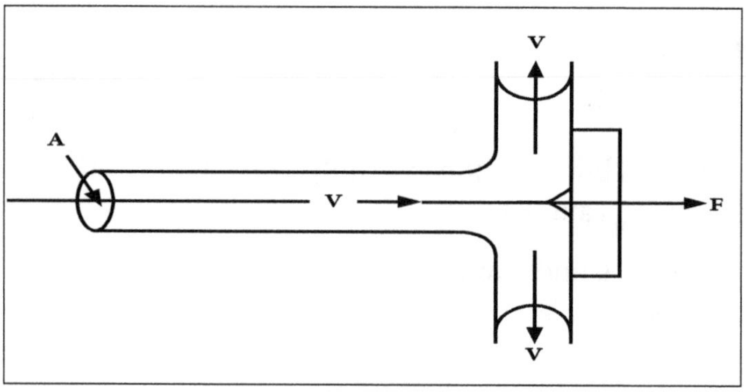

Considering only forces in a horizontal direction $u_1 = V$ and $u_2 = 0$ therefore,

$$F = Q\rho V = \rho A V^2$$

Jet force on a flat plate at an angle θ:

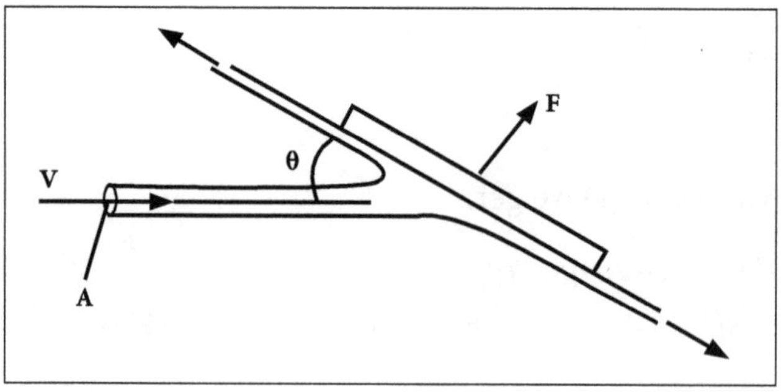

Considering only forces Normal to plate surface $u_1 = V \sin\theta$ and $u_2 = 0$ therefore,

$$F = Q\rho V \sin\theta = \rho A V^2 \sin\theta$$

Where, θ = 900 then, F = ρAV2 as above.

Jet force on angled plate (θ < 900):

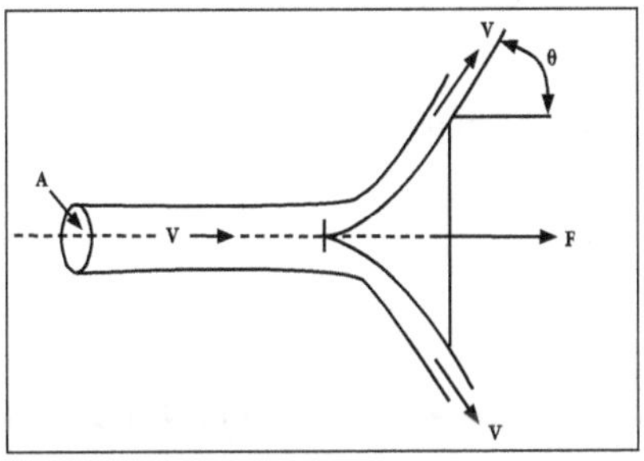

Considering only forces in a horizontal direction u_1 = V and u_2 =V cos θ therefore,

$$F = Q\rho V(1-\cos\theta) = \rho AV^2(1-\cos\theta)$$

Jet force on angled plate (θ > 900):

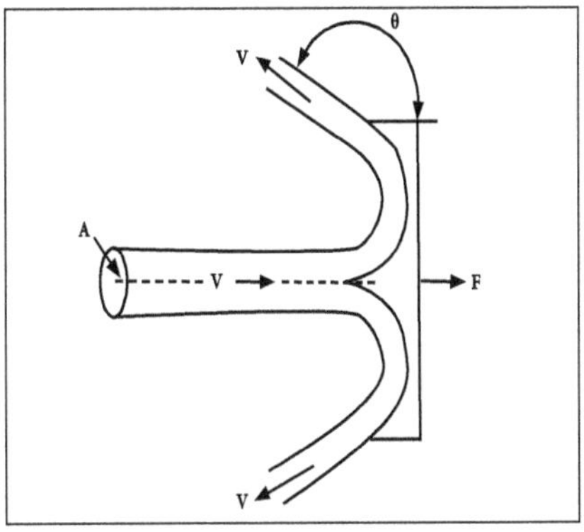

Considering only forces in a horizontal direction u_1 = V and u_2 =V cos θ therefore,

$$F = Q\rho V(1-\cos\theta) = \rho AV^2(1-\cos\theta)$$

Jet force on angled plate (θ = 1800):

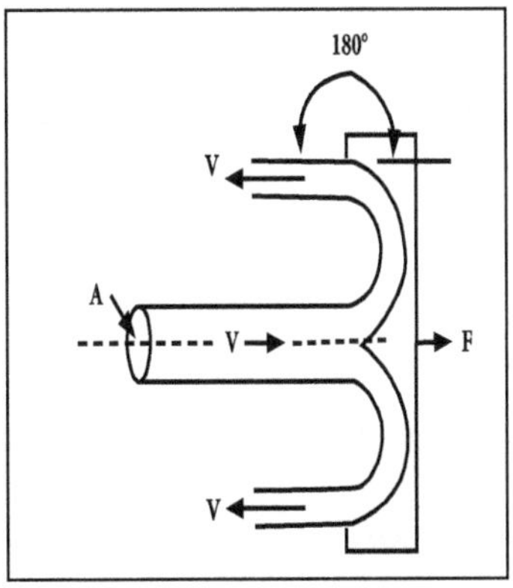

$$u_1 = V \text{ and } u_2 = V\cos\theta \text{ therefore,}$$

$$F = Q\rho V(1-\cos 180°) = 2Q\rho V = 2\rho AV^2$$

4.1.2 Jet Forces on Moving Plate

Jet Force on a Moving Flat Plate:

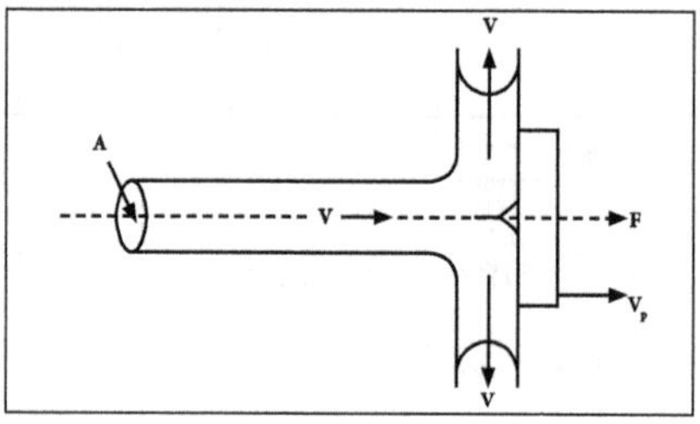

Considering only forces in a horizontal direction:

$$u_1 = V \text{ and } u_2 = V_p$$

And let, $r = V_p / V$ therefore,

$$F = Q\rho(V - V_p) = \rho AV(V - V_p)$$

And,

$$F = \rho AV^2(1-r)$$

The power (P) generated by the force on the moving plate = $P = F \cdot V_p$.

Jet force on angled moving plate:

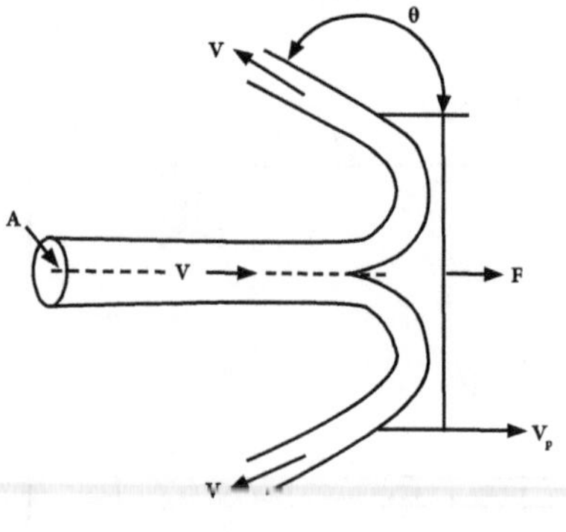

Considering only forces in a horizontal direction:

$$u_1 = V \text{ and } u_2 = V_p + (V - V_p)\cos\theta$$

And let $r = V_p / V$ therefore,

$$F = \rho AV(V - V_p)(1 - \cos\theta) = \rho AV^2(1 - r)(1 - \cos\theta)$$

The power (P) generated by the force on the moving plate $P = F \cdot V_p$.

4.1.3 Jet Forces on Vanes

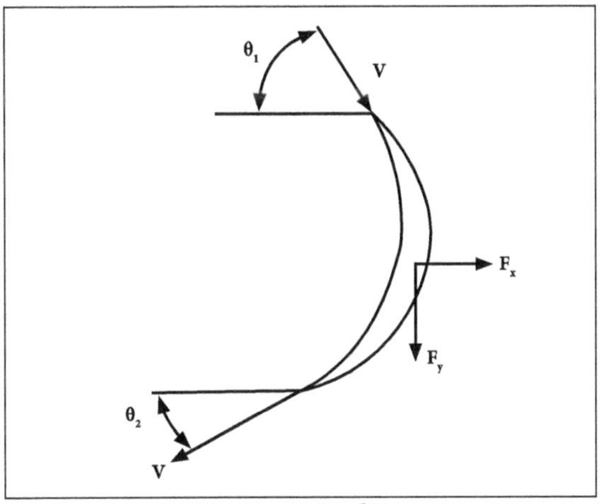

Jet Force on Fixed Vanes

In the x Direction: $u_{1x} = V\cos\theta_1 \cdot u_{2x} = -V\cos\theta_2$

$$F_x = Q\rho V(\cos\theta_1 + \cos\theta_2) = \rho AV^2(\cos\theta_1 + \cos\theta_2)$$

In the y Direction: $u_{1y} = V\sin\theta_1\ u_{2y} = V\sin\theta_2$

$$F_y = Q\rho V(\sin\theta_1 - \sin\theta_2) = \rho AV^2(\sin\theta_1 - \sin\theta_2)$$

Jet Force on Moving Vanes:

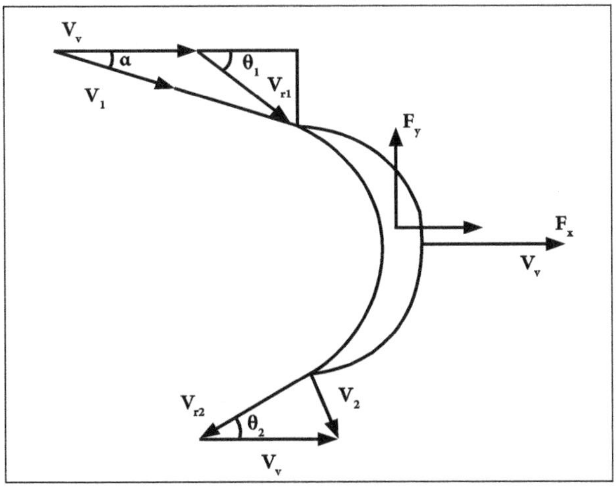

In the x Direction:

$$u_{1x} = V_1 \cos\alpha$$

$$u_{2x} = V_v - V_{r2} \cos\theta_2 \ldots (V_{r2} = V_{r1} = V_1 \sin\alpha / \sin\theta_1)$$

$$F_x = Q\rho V_1 (\cos\alpha + [\sin\alpha / \sin\theta_1] \cos\theta_2 - r >)$$

$$r = V_v / V_1$$

In the y Direction:

$$u_{1y} = V_1 \sin\alpha$$

$$u_{2y} = V_{r2} \sin\theta_2 \ldots (V_{r2} = V_{r1} = V_1 \sin\alpha / \sin\theta_1)$$

$$\text{In the y direction } F_y = Q\rho V_1 \sin\alpha (1 - \sin\theta_2 / \sin\theta_1)$$

In the vane is moving in the x direction the power developed by the vane $P = F_x \cdot V_V$.

Jet Force on Curved Vanes

When a jet hits a curved vane, it arrives on the vane at the same angle as the vane. The jet is diverted with no splashing by the curve of the vane. If there is no friction, the direction of the jet is changed and not its speed.

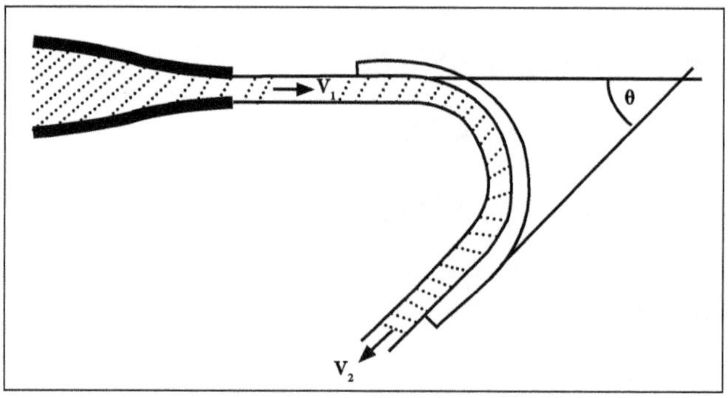

If the deflection angle is θ then the impulsive force is:

$$F = m'\Delta v = m'v_1 \{2(1 - \cos\theta)\}^{1/2}$$

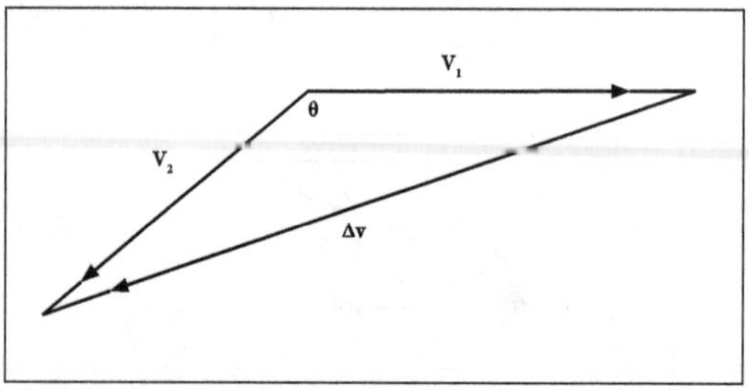

The direction of the force on the fluid is in the direction of Δv and the direction of the force on the vane is opposite. The force may be resolved to find the forces acting horizontally and/or vertically.

It is often necessary to solve the horizontal force and this is done as follows:

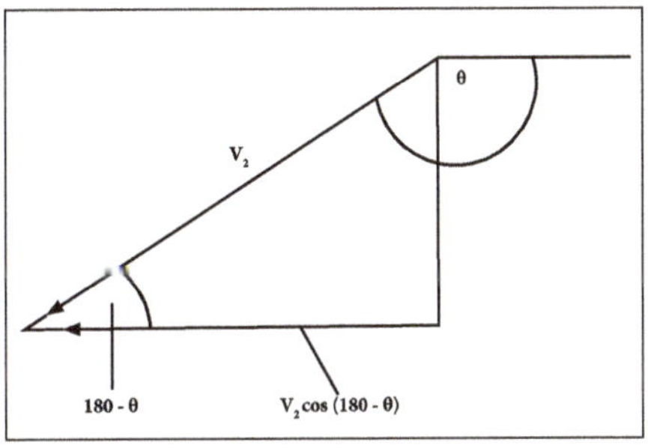

Initial horizontal velocity $=v_{H1}=v_1$

Final horizontal velocity $=v_{H2}=-v_2\cos(180-\theta)=v_2\cos\theta$

Change in horizontal velocity $=\Delta v_{H1}$

Since, v_2-v_1 this becomes $\Delta v_h=\{v_2\cos\theta-v_1\}=v_1\{\cos\theta-1\}$

Horizontal force on fluid $=m'v_1\{\cos\theta-1\}$

The horizontal force on the vane is opposite so,

Horizontal force $=m\Delta v_H=m'v_1\{1-\cos\theta\}$

4.2 Velocity Diagrams, Work Done and Efficiency

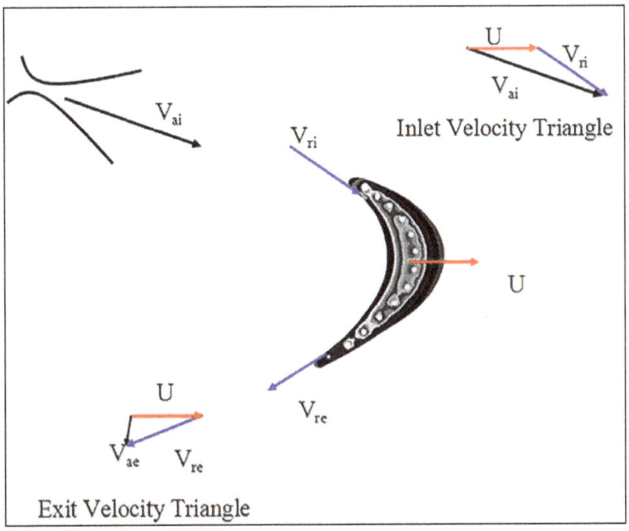

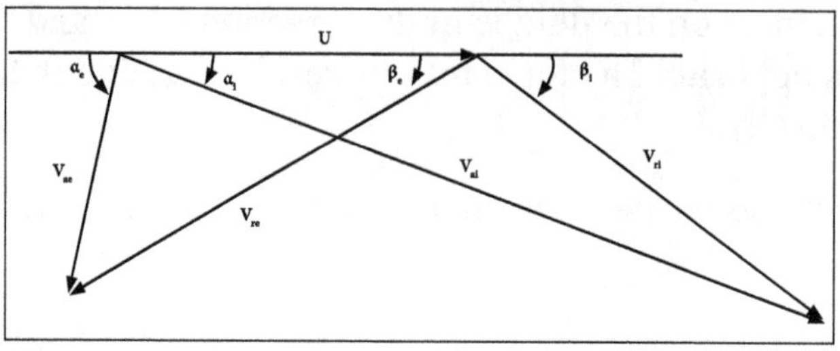

- V_{ai} : Inlet Absolute Velocity.
- V_{ri} : Inlet Relative Velocity.
- V_{re} : Exit Relative Velocity.
- V_{ae} : Exit Absolute Velocity.
- ai : Inlet Nozzle Angle.
- bi : Inlet Blade Angle.
- be : Exit Blade Angle.
- ai : Exit Nozzle Angle.

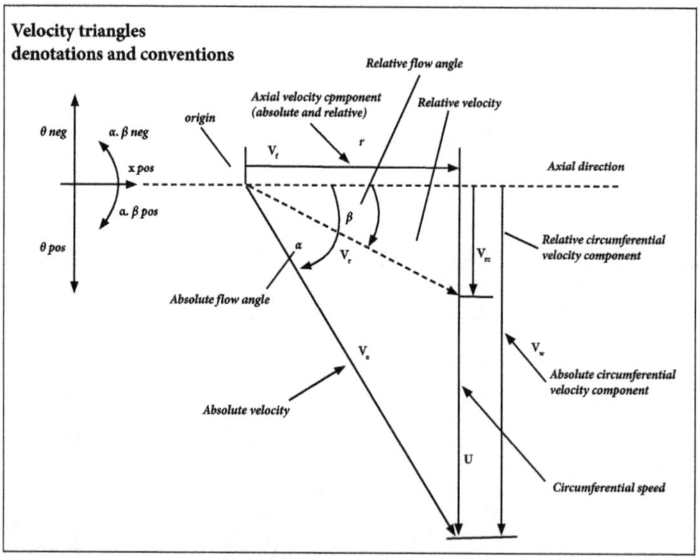

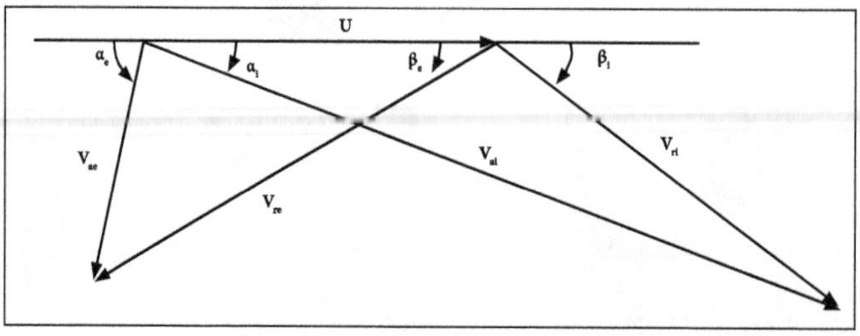

- The stream is delivered to the wheel at an angle ai and velocity V_{ai}.
- An increase in ai, reduces the value of useful component (Absolute circumferential Component).
- Inlet Whirl Velocity, $V_{wi} = V_{ai}\cos(ai)$.
- An increase in ai, increases the value of axial component also called as flow component.
- Flow component $V_{fi} = V_{ai}\sin(a_i) = V_{ri}\sin(bi)$.

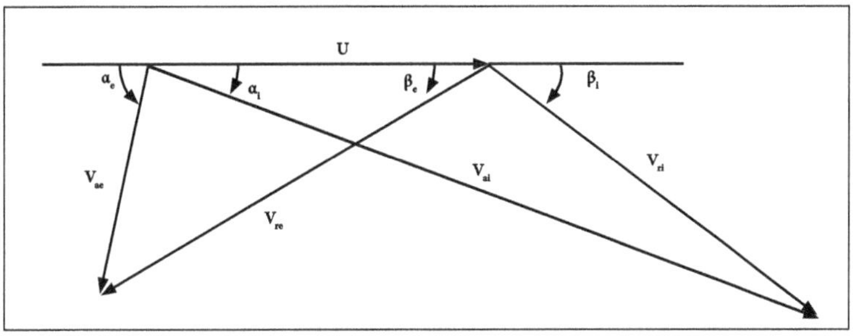

- If the stream is to enter and leave the blades without shock or much loss, then relative velocity should be tangential to the blade inlet tip.
- V_{ri} – Inlet blade angle.
- V_{re} – Exit blade angle.
- A blade is said to be symmetric if bi = be.
- The flow velocities between two successive blades at inlet and exit are V_{fi} & V_{fe}.
- The axial (basic useful) components or whirl velocities at inlet and exit are V_{wi} & V_{we}.

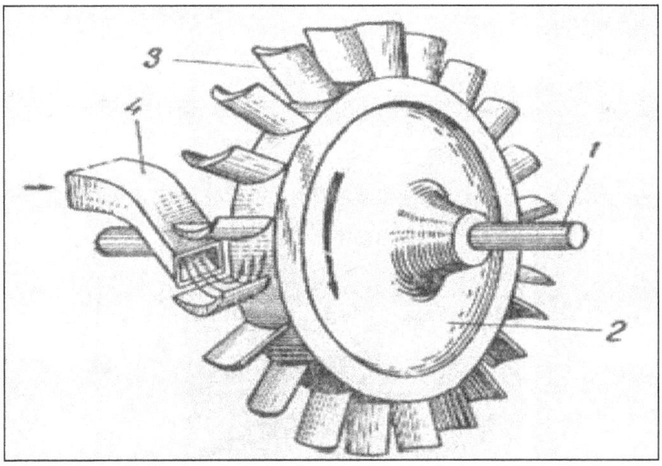

Impulse Turbine.

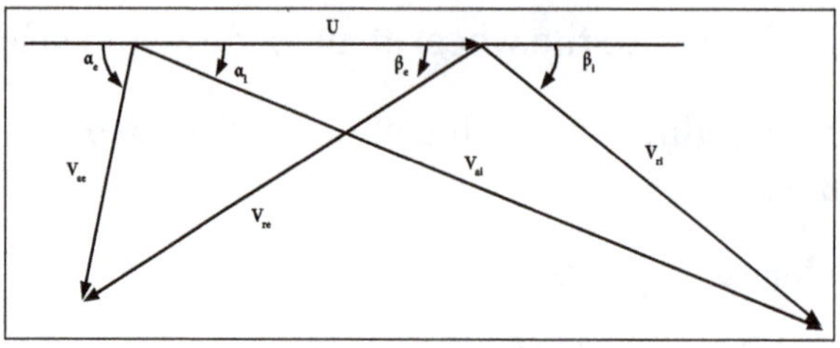

Newton's Second Law for an Impulse Blade

The tangential force acting of the jet is:

F = Mass flow rate × Change of velocity in the tangential direction

Tangential relative velocity at blade Inlet: $V_{ri}\cos(\beta_i)$

Tangential relative velocity at blade exit: $-V_{re}\cos(\beta_e)$

Change in velocity in tangential direction: $-V_{re}\cos(\beta_e) - V_{ri}\cos(\beta_i)$

$$-(V_{re}\cos(\beta_e) + V_{ri}\cos(\beta_i))$$

Tangential force, $F_A = -\dot{m}(V_{re}\cos\beta_e + V_{ri}\cos\beta_i)$

The reaction to this force provides the driving thrust on the wheel.

The driving force of wheel:

$$F_R = \dot{m}(V_{re}\cos\beta_e + V_{ri}\cos\beta_i)$$

Power output of the blade:

$$P_b = \dot{m}U(V_{re}\cos\beta_e + V_{ri}\cos\beta_i)$$

Diagram efficiency or Blade efficiency:

$$\eta_d = \frac{\text{Power output}}{\text{kinetic power of inlet steam}}$$

$$\eta_d = \frac{\dot{m}U(V_{re}\cos\beta_e + V_{ri}\cos\beta_i)}{\dot{m}V^2_{ai}}$$

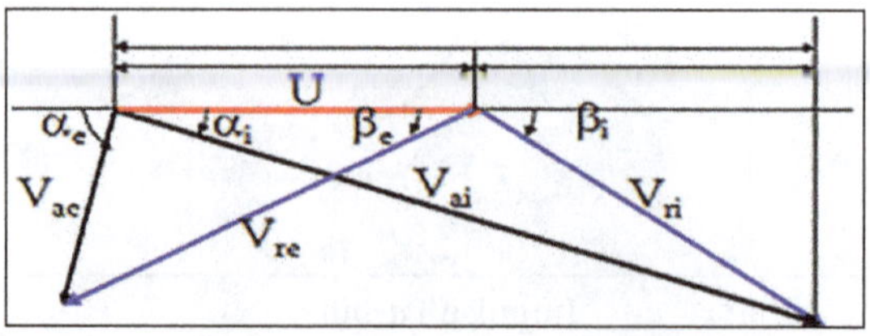

$$\eta_d = \frac{2U(kV_{ri}\cos\beta_e + V_{ri}\cos\beta_i)}{V_{ai}^2}$$

$$\eta_d = \frac{2UV_{ri}(k\cos\beta_e + \cos\beta_i)}{V_{ai}^i}$$

$$V_{ai}\cos\alpha_i = U + V_{ri}\cos\beta_i$$

$$V_{ri} = \frac{V_{ai}\cos\alpha_i - U}{\cos\beta_i}$$

$$\eta_d = \frac{2U(V_{ai}\cos\alpha_i - U)\left(k\frac{\cos\beta_e}{\cos\beta_i} + 1\right)}{V_{ai}^2}$$

$$\eta_d = \frac{2U(kV_{ri}\cos\beta_e + V_{ri}\cos\beta_i)}{V_{ai}^2}$$

$$\eta_d = \frac{2UV_{ri}(k\cos\beta_e + \cos\beta_i)}{V_{ai}^2}$$

$$V_{ai}\cos\alpha_i = U + V_{ri}\cos\beta_i$$

$$V_{ri} = \frac{V_{ai}\cos\alpha_i - U}{\cos\beta_i}$$

$$\eta_d = \frac{2U(V_{ai}\cos\alpha_i - U)\left(k\frac{\cos\beta_e}{\cos\beta_i} + 1\right)}{V_{ai}^2}$$

$$\eta_d = 2\left\{\frac{U}{V_{ai}}\cos\alpha_i - \left[\frac{U}{V_{ai}}\right]^2\right\}\left\{k\frac{\cos\beta}{\cos\beta_{i_e}} + 1\right\}$$

Blade Speed Ratio

$$\eta_d = 2\phi\{\cos\alpha_i - \phi\}\left\{\left(k\frac{\cos\beta_e}{\cos\beta_i} + 1\right)\right\}$$

For a given shape of the blade, the efficiency is a strong function of ϕ.

For maximum efficiency: $\frac{d\eta_d}{d\phi} = 0$

$$2\{\cos\alpha_i - 2\phi\}\left\{\left(k\frac{\cos\beta_e}{\cos\beta_i} + 1\right)\right\} = 0$$

$$\{\cos\alpha_i - 2\phi\} = 0 \Rightarrow \phi = \frac{\cos\alpha_i}{2}$$

$$\eta_{d,max} = 2\cos\alpha_i\left\{\cos\alpha_i - \frac{\cos\alpha_i}{2}\right\}\left\{\left(k\frac{\cos\beta_e}{\cos\beta_i} + 1\right)\right\}$$

$$\eta_{d,max} = \phi^2\cos^2\alpha_i\left\{\left(k\frac{\cos\beta_e}{\cos\beta_i} + 1\right)\right\}$$

Impulse Reaction Turbine

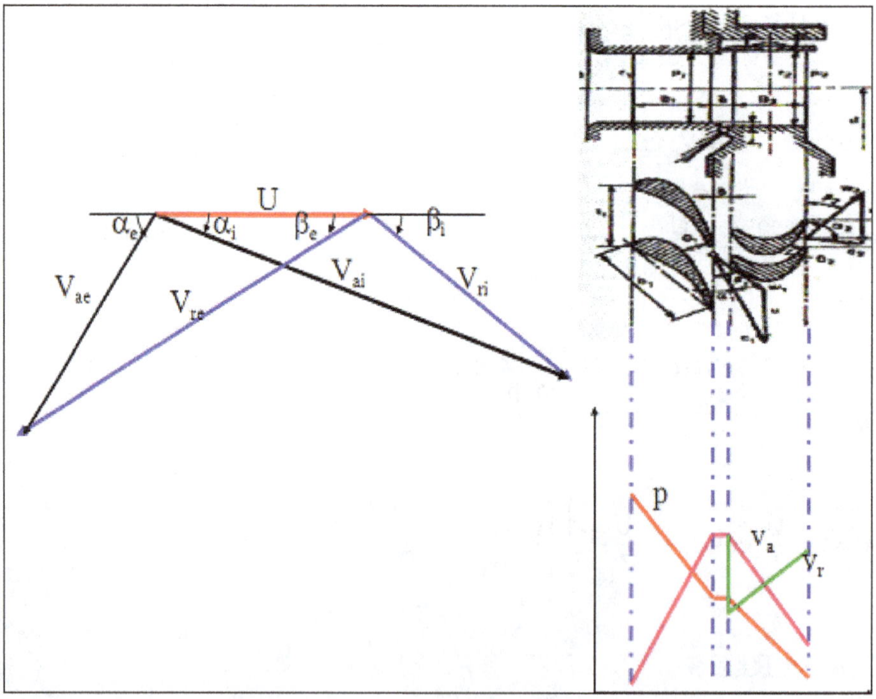

The reaction effect is an addition to impulse effect.

The degree of reaction:

$$\Delta = \frac{\text{The enthalpy drop in the moving blades}}{\text{The enthalpy drop in the stage}}$$

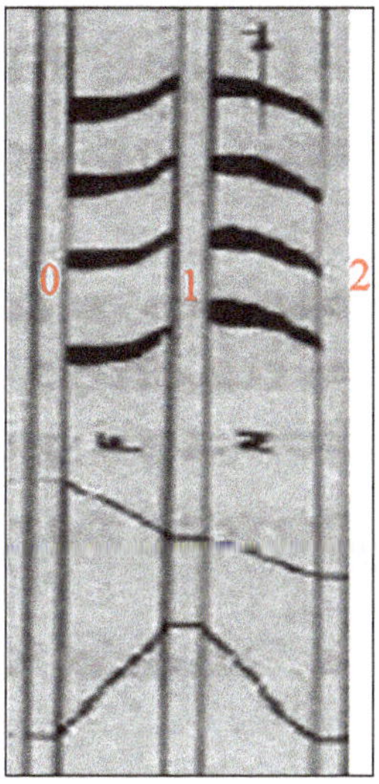

First law for fixed blades:

$$h_o - h_1 = \frac{V_1^2 - V_o^2}{0}$$

First law for moving blades:

$$h_1 - h_2 = \frac{V_{r^2}^2 - V_{r1}^2}{2}$$

$$h_o - h_2 = \frac{V_1^2 - V_o^2}{2} + \frac{V_{r2}^2 - V_{r1}^2}{2}$$

$$h_o - h_2 = \frac{V_{a1}^2 - V_o^2}{2} + \frac{V_{r2}^2 - V_{r1}^2}{2}$$

$$\Delta = \frac{\text{The enthalpy drop in the moving blades}}{\text{The enthalpy drop in the stage}}$$

$$\Delta = \frac{h_1 - h_2}{h_o - h_2}$$

$$\Delta = \frac{h_1 - h_2}{h_o - h_2} = \frac{V_{r2}^2 - V_{r1}^2}{V_{a1}^2 - V_o^2 + V_{r2}^2 - V_{r1}^2}$$

$$V_{r2}^2 = V_{r1}^2 + \left(\frac{\Delta}{1-\Delta}\right)\left\{V_{a1}^2 - V_o^2\right\}$$

$$V_{r2} = \sqrt{V_{r1}^2 + \left(\frac{\Delta}{1-\Delta}\right)\left\{V_{a1}^2 - V_o^2\right\}}$$

$$\eta_d = \frac{2U\left(V_{r2}\cos\beta_2 + V_{r1}\cos\beta_1\right)}{V_{a1}^2}$$

$$V_{r2} = \sqrt{V_{r1}^2 + \left(\frac{\Delta}{1-\Delta}\right)\left\{V_{a1}^2 - V_o^2\right\}}$$

$$V_{a1}\cos\alpha_1 = U + V_{r1}\cos\beta_1 \qquad V_{r1} = \frac{V_{a1}\cos\alpha_1 - U}{\cos\beta_1}$$

Losses in nozzle, Nozzle blade loss factor, φ:

$$\phi = \frac{\text{Actual absolute inlet velocity}}{\text{Isoentropic Velocity at nozzle exit}} = \frac{V_{a1}}{V_{n,iso}}$$

$$\eta_{stage} = \frac{2U\left(V_{r2}\cos\beta_2 + V_{r1}\cos\beta_1\right)}{V_{n,iso}^2}$$

$$\eta_{stage} = \frac{2U\phi^2\left(\sqrt{V_{r1}^2 + \left(\frac{\Delta}{1-\Delta}\right)\left\{V_{a1}^2 - V_o^2\right\}}\cos\beta_2 + V_{r1}\cos\beta_1\right)}{V_{a1}^2}$$

$$V_{r1} = \frac{V_{a1}\cos\alpha_1 - U}{\cos\beta_1}$$

$$\eta_{stage} = \frac{2U\phi^2\left(\sqrt{\left(\frac{V_{a1}\cos\alpha_1 - U}{\cos\beta_1}\right)^2 + \left(\frac{\Delta}{1-\Delta}\right)\left\{V_{a1}^2 - V_o^2\right\}}\cos\beta_2 + \left(\frac{V_{a1}\cos\alpha_1 - U}{\cos\beta_1}\right)\cos\beta_1\right)}{V_{a1}^2}$$

$$\eta_{stage} = \frac{2\frac{U}{V_{a1}}\phi^2\left(\sqrt{\left(\frac{\sqrt{\cos\alpha_1 - \frac{U}{V_{a1}}}}{\cos\beta_1}\right)^2 + \left(\frac{\Delta}{1-\Delta}\right)\left\{1-\left(\frac{V_o}{V_{a1}}\right)^2\right\}}\cos\beta_2 + \left(\frac{\cos\alpha_1 - \frac{U}{V_{a1}}}{\cos\beta_1}\right)\cos\beta_1\right)}{V_{a1}^2}$$

For a given shape of the blade, the efficiency is a strong function of U/V_{a1}.

For maximum efficiency: $\dfrac{d\eta_{stage}}{d\left(\dfrac{U}{V_{a1}}\right)} = 0$

Centrifugal and Reciprocating Pumps

5.1 Centrifugal Pumps: Classification and Working

Centrifugal pump uses the centrifugal force to pump the fluid to a certain head. The centrifugal pump is classified based on:

Working of Centrifugal Pump

Priming is the first operation in which the suction pipe, casing of the pump and the portion of the delivery pipe up to delivery valve are completely filled with liquid which is to be pumped so that all the air or gas or vapor from this portion of the pump is driven out and no air pocket is left.

The delivery valve being in closed position, the electric motor is started to rotate the impeller. When the delivery valve is opened the liquid is made to flow in an outward radial direction thereby having the vanes of the impeller, at the outer circumference with high velocity and pressure and thereby at the eye of the impeller vacuum is formed which causes for the water to enter the casing or impeller from the sump. Suitable casing is provided in order to convert velocity energy into pressure energy with gradual reduction of velocity so that loss of energy due to eddy formation is minimum.

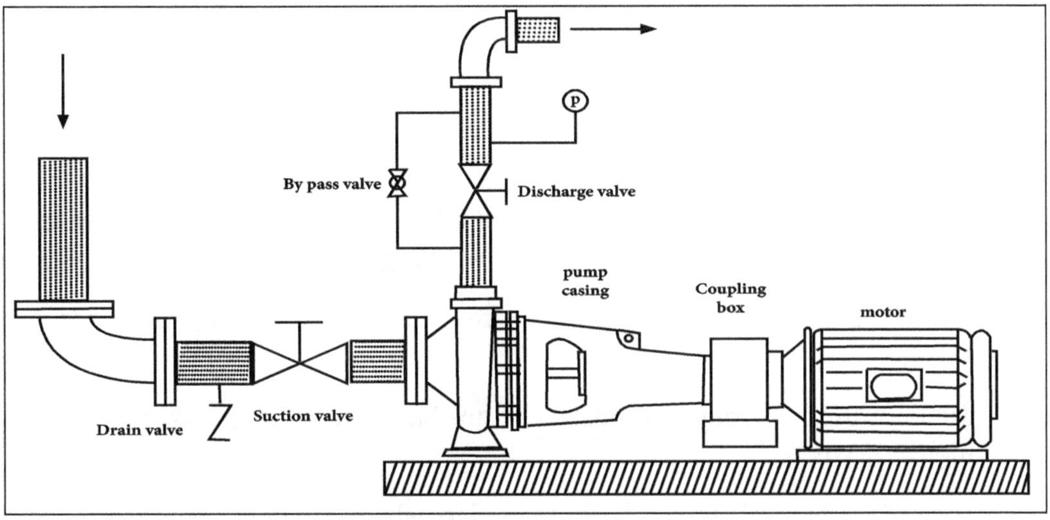

Centrifugal pump system.

1. Type of casing According to the type of casing centrifugal pump have two basic types:

- Volute casing: In order to increase the pressure head, the velocity of fluid needs to decrease. It is done by gradually increasing the area of the casing from the impeller out let known as volute casing.

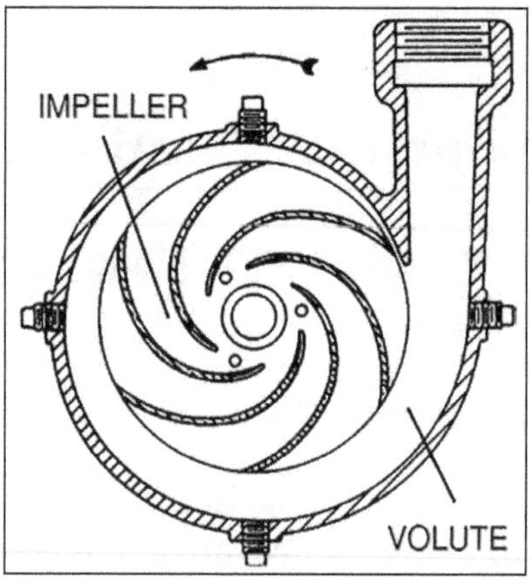

Volute casing.

- Vortex Casing: If a circular chamber known as vortex is introduce between the impeller and chamber then the casing is called vortex casing. Its main function is to convert the kinetic energy into pressure energy.

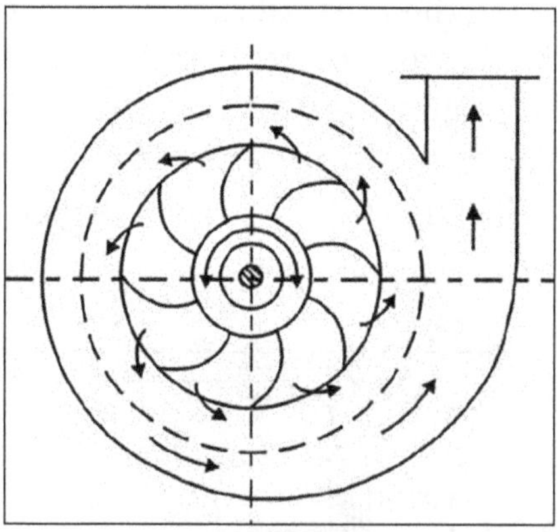
Vortex casing.

2. Working head: Each pump can pump up a fluid to a certain height so following are the types of centrifugal pump according to the working head:

- Low lift centrifugal pump: They pump are capable of working against heads up to 15 m.

- Medium lift centrifugal pumps: They are basically used against the heads as high as 40 m.

- High lift centrifugal pumps: They are used to deliver liquids at heads above 40 m.

3. Liquid handled according to the type of liquid pumped centrifugal pump is classified into three types:

- Pure liquid: When pure liquid is to pump the centrifugal pump with the closed impeller are used because they have better guidance and high efficiency.
- Little impure liquid: When liquid have a little impurity then centrifugal pump with semi open impeller are used.
- Liquid with solid matter: When sewage, paper pulp, water containing sand or grit is to be pumped then pump with open impeller is used.

4. Number of impellers per shaft The following are the classification based on the number of pumps:

- Single stage centrifugal pump: It has only one impeller attached to the shaft. They are used in place where low head and low discharge rate is required.
- Multi-stage centrifugal pump: It has more than one impeller attached to a single shaft. They are used in place where high head and high discharge is required.

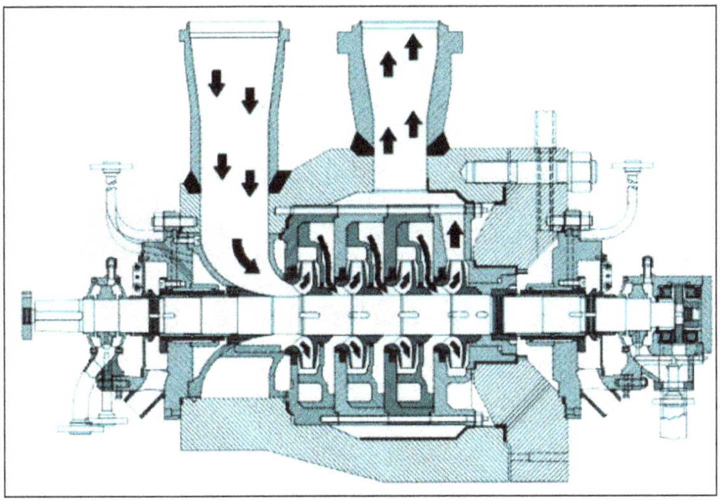

Multi-stage centrifugal pump.

5. Number of entrances to the impeller: Usually it is consider that pump have only one entrance and one leaving point but there could be more than one:

- Single entry or single suction pump: They also called single suction pump is one in which water is admitted on one side of the impeller.
- Double entry or double suction pump: They are the one in which water is admitted on both sides of the impeller, axial thrust is neutralized. They are usually employed for pumping large quantities of fluid.

6. Relative direction of flow through impeller depending upon the direction of the liquid flow centrifugal pump can be classified as follow:

Radial Flow Pump:

In a radial flow pump, the liquid enters at the center of the impeller and is directed out along the impeller blades in a direction at right angles to the pump shaft. The impeller of a typical radial flow pump and the flow through a radial flow pump are shown below in figure.

They are the one in which regular radial flow impellers are used.

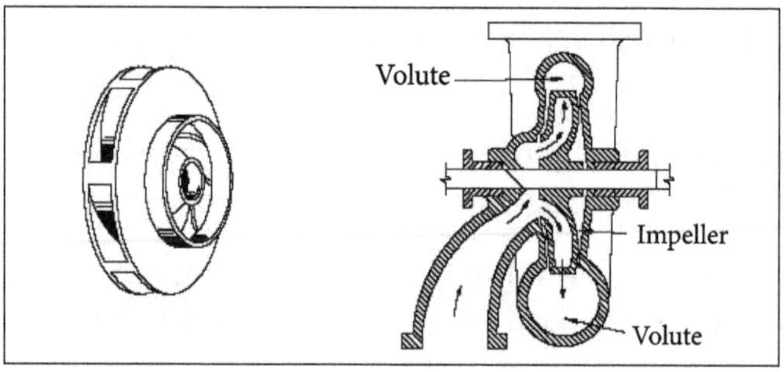

Axial Flow Pump:

In an axial flow pump, the impeller pushes the liquid in a direction parallel to the pump shaft. Axial flow pumps are sometimes called propeller pumps because they operate essentially the same as the propeller of a boat. The impeller of a typical axial flow pump and the flow through a radial flow pump are shown in figure.

They are the one in which designed to deliver low head but huge quantity of water.

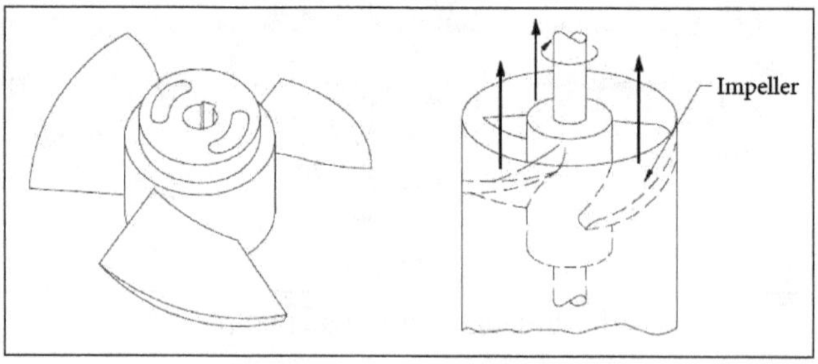

Mixed Flow Pump:

The mixed flow centrifugal pumps borrow characteristics from both radial flow and axial flow pumps. As liquid flows through the impeller of a mixed flow pump, the impeller blades push the liquid out away from the pump shaft and to the pump suction at an angle greater than 90 degrees.

The impeller of a typical mixed flow pump and the flow through a mixed flow pump are shown in figure.

Mixed flow pump are mostly used for irrigation purpose.

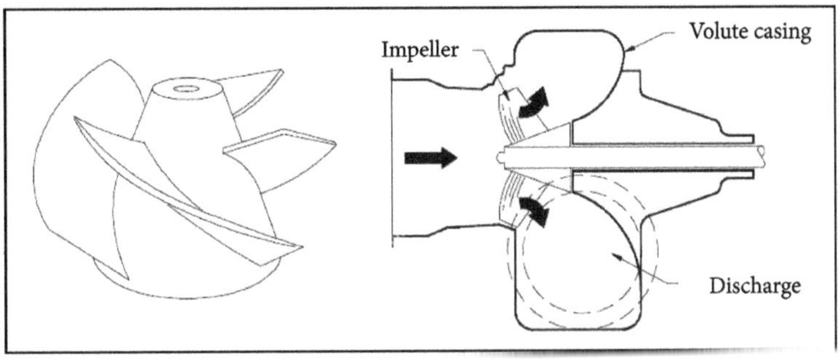

Problems

1. A centrifugal pump delivers water against a net head of 14.5 meters and a design speed of 1000 rpm. The vanes are carved back to an angle of 30° with the periphery. The impeller diameter is 300 mm and outlet width 50 mm. Let us determine the discharge of the pump if Manometric efficiency is 95%.

Solution:

Given:

- Net head, $H_m = 14.5m$
- Speed $N = 1000$ r.p.m
- Vane angle at outlet $\varphi = 30°$

Impeller diameter means the diameter of the impeller at outlet.

$\therefore$ Diameter $D_2 = 300mm = 0.30m$

Outlet width $B_2 = 50mm = 0.05m$

Manometric efficiency, $\eta_{man} = 95\% = 0.95$

To find: The Discharge of the Pump.

Tangential velocity of impeller at outlet:

$$u_2 = \frac{\pi D_2 N}{60} = \frac{\pi \times 0.30 \times 1000}{60} = 15.70\, m/s$$

Now, using equation $\eta_{man} = \dfrac{gH_m}{V_{w_2} \times u_2}$

$$\therefore 0.95 = \frac{9.81 \times 14.5}{V_{w_2} \times 15.70}$$

$$\therefore V_{w_2} = \frac{0.95 \times 14.5}{0.95 \times 15.70} = 9.54\,m/s$$

From outlet velocity triangle, we have:

$$\tan\phi = \frac{V_{f_2}}{(u_2 - V_{w_2})} \text{ or } \tan 30^\circ = \frac{V_{f_2}}{(15.70 - 9.54)} = \frac{V_{f_2}}{6.16}$$

$$\therefore V_{f_2} = 6.16 \times \tan 30^\circ = 3.556\,m/s$$

Discharge $\varphi = \pi \times D_2 \times B_2 \times V_{f_2}$

$$= \pi \times 0.30 \times 0.55 \times 3.55b$$

$$= 0.1675\ m^3/s$$

Table: The head - discharge characteristics of a centrifugal pump is given below:

Discharge- (lit/sec)	0	10	20	30	40	50
Head (meters)	25.3	25.5	24.5	22.2	18.7	12.0

2. The pump delivers fresh water through a 500 m long, 15 cm diameter pipe line having friction coefficient of f = 0.025. The static lift is 15 m. Neglecting minor losses in the pipe flow; let us determine is: (i) the discharge of the pump under the above conditions (ii) driving power of the pump motor. Assume a pump efficiency of 72%.

Solution:

Given data:

- $\eta_o = 80\% = 0.8$
- $P = 136\,kw$
- $H = 16m$
- $N = 120$ mpm
- Hydraulic losses = 15% of available energy

- Assuming discharge of outlet = Radial

$$V_{w2} = 0, \; V_{r1} = V_{f1}$$

To find:

(i) The Discharge of the Pump under the Above Conditions:

Peripheral velocity of wheel:

$$m = 3.3\sqrt{H} = 3.3\sqrt{16}$$
$$m = 13.2\,m/s$$

Radial Velocity of flow:

$$V_{r1} = 1.1\sqrt{H} = 1.1\sqrt{16}$$
$$V_{f1} = 4.4\,m/s$$

Hydraulic efficiency:

$$\mu_h = \frac{\text{Head at inlet} - \text{Hydraulic losses}}{\text{Head at inlet}}$$

(ii) Driving Power of the Pump Motor:

$$\mu_H = \frac{H - 0.15H}{H} = \frac{0.85H}{H} = 0.85 = 85\%$$
$$\mu_H = 85\%$$

$$\mu_H = \frac{V_{\omega 1} u_1}{gH}$$
$$0.85 = \frac{V_{\omega 1} \times 13.2}{9.81 \times 16}$$
$$V_{\omega 1} = 10.107\ m/s$$

Guide blade angle:

$$\tan 2 = \frac{V_{r1}}{V_{\omega 1}} = \frac{4.4}{10.10^7} = 0.4353$$
$$\alpha = \tan^{-1}(0.h353)$$
$$\alpha = 23.523$$

Vane angle at inlet:

$$\tan\theta = \frac{V_{f1}}{u_1 - V_{\omega 1}} = \frac{4.4}{13.2 - 10.107} = 1.422$$

$$\theta = \tan^{-1}(1.422)$$

$$\theta = 54.89^\circ$$

Velocity:

$$u_1 = \frac{\pi D_1 N}{60}$$

$$13.2 = \frac{\pi \times D_1 \times 120}{60}$$

$$D_1 = 2.100\,\text{m}$$

Overall efficiency:

$$\mu_o = \frac{P}{wQ_H}$$

$$0.8 = \frac{136}{9.81 \times Q \times 16}$$

$$Q = 1.0830\,\text{m}^3/\text{s}$$

$$Q = \pi D_1 B_1 V_{f1}$$

$$1.830 = \pi \times 2.100 \times B_1 \times h.h$$

$$B_1 = 0.0373\,\text{m}$$

5.2 Work Done, Manometric Head and Specific Speed

Expression for Work Done

In centrifugal pump the liquid enters the impeller at its center and leaves at its outer periphery.

Assumptions

- Liquid enters the impeller eye in radial direction. The whirl component $V_{w1} = 0$.

  $V_{f_1} = V_1$, $a = 90^\circ$

- No loss of energy in the impeller due to friction and eddy formation.

- No loss at entry due to shock.
- There is uniform velocity distribution in the narrow passages formed between two adjacent vanes.

Let,

D_1 – Diameter of the impeller at inlet.

N = Speed of the impeller in rpm.

$$\omega = \text{Angular velocity} = \left(\frac{2\pi N}{60}\right) \text{rad/s}$$

u = Tangential velocity of the impeller at inlet

$$u_1 = \frac{\pi D_1 N}{60} = \frac{2\pi R_1 N}{60} = \omega \cdot R_1$$

Similarly,

$$u_2 = \omega \cdot R_2$$

V_1 = Absolute velocity of water at inlet.

V_{w_1} = Velocity of whirl at inlet.

V_{r_1} = Relative velocity of liquid at inlet.

V_{f_1} = Velocity of flow at inlet.

α = Angle made by absolute velocity (V_1) at inlet with the direction of motion of vane.

θ = Angle made by the relative velocity (V_{r1}) at inlet with the direction of motion of vane.

V_2, V_{r_2}, V_{f_2}, β, ϕ are corresponding values at outlet.

While passing through the impeller, the velocity of whirl charges and there is a change of momentum.

Torque on the impeller = Rate of change of moment of momentum

Moment of momentum at inlet = o (V_{w1} = o)

$$\text{At outlet} = \frac{W}{g}\left(V_{w_2} \cdot R_2\right)$$

$$\text{Torque} = \frac{W}{g}\left(V_{w_2} \cdot R_2\right)$$

Work done per second = Torque × Angular velocity

$$= \frac{W}{g} V_{w_2} \cdot R_2 \times \omega$$

$$= \frac{W}{g} \cdot V_{w_2} \cdot u_2 \quad (\because u_2 = \omega \cdot R_2)$$

$$\text{Work done per second per unit weight of liquid} = \frac{V_{w_2} \cdot u_2}{g}$$

Previous equation has been developed assuming flow at inlet to be radial. If the flow is not radial the expression for work done may be written as:

$$\text{Work done per second} = \frac{W}{g}\left[V_{w_2} \cdot u_2 - V_{w_1} \cdot u_1\right]$$

$$\text{Work done per second per unit weight of liquid} = \frac{1}{g}\left[V_{w_2} \cdot u_2 - V_{w_1} \cdot u_1\right]$$

This is known as Euler Momentum equation for centrifugal pump.

5.2.1 Manometric Head

Heads on a Centrifugal Pump

- Suction head ( ): The vertical distance between the liquid level in the sump and the centre line of the pump is called suction head. It is expressed as meters.
- Delivery head (h_d): The vertical distance between the centre line of the pump and the liquid level in the overhead tank or the supply point is called delivery head. It is expressed in meters.
- Static head (H_S): The vertical difference between the liquid levels in the overhead tank and the sump, when the pump is not working is called static head. It is expressed as meters.

  $$H_S = \left(h_s + h_d\right)$$

- Friction head (h_f): The sum of the head loss due to the friction in the suction and delivery pipes is called friction head. The friction loss in both the pipes is calculated using the Darcy's equation.

$$h_f = \left(fLV^2/2gD\right)$$

- Total head (H): The sum of the static head Hs, friction head (hf) and the velocity head in the delivery pipe (Vd2/2g) is called total head. Where, Vd= Velocity in the delivery pipe.

$$\therefore H_m = \left(h_s + h_d + h_f + \frac{Vd^2}{2g}\right)$$

- Manometric head (H_m): The total head developed by the pump is called manometric head. This head is slightly less than the head generated by the impeller due to some losses in the pump.

$$\therefore H_m = H + \frac{Vs^2}{2g} - \frac{Vd^2}{2g}$$

Problems

1. Let us calculate the vane angle at the inlet of a centrifugal pump impeller having 200 mm diameter at inlet and 400 mm diameter at outlet. The impeller vanes are set back at angle of 45° to the outer rim and the entry of the pump is radial. The pump runs at 1,000 rpm and the velocity of flow through the impeller is constant at 3 m/s. Also calculate the work done per kN of water and the velocity as well as direction of the water at outlet.

Solution:

Given data:

- $D_1 = 0.2$
- $D_2 = 0.4$ m
- $\varphi = 45°$

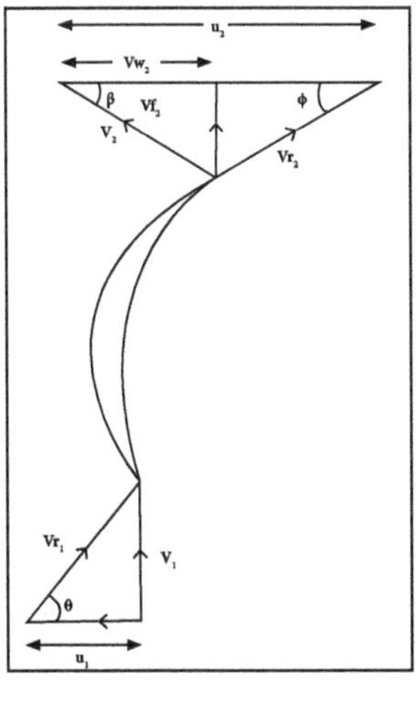

Entry of the pump is radical, $\alpha = 90°$.

$$N = 1000 \text{ rpm}$$

Velocity of flow through the impeller is constant.

$$Vf_1 = Vf_2 = 3 \text{ m/s}$$

$$u_1 = \frac{\pi D_1 N}{60}$$

$$u_1 = \frac{\pi \times 0.2 \times 1000}{60}$$

$$u_1 = 10.47 \text{ m/s}$$

$$u_2 = \frac{\pi D_2 N}{60}$$

$$u_2 = \frac{\pi \times 0.4 \times 1000}{60}$$

$$= 20.94 \text{ m/s}$$

For inlet velocity triangle calculating vane angle at inlet θ.

$$\tan\theta = \frac{Vf_1}{u_1}$$

$$\tan\theta = \frac{3}{10.47}$$

$$\theta = 15.99^\circ$$

Calculating work done per kN of water $= \frac{1}{g} Vw_2 u_2$

Calculating V_{w2} from outlet velocity triangle:

$$\tan\phi = \frac{Vf_2}{u_2 - Vw_2}$$

$$\tan 45^\circ = \frac{3}{20.94 - Vw_2}$$

$$20.94 - Vw_2 = \frac{3}{\tan 45^\circ}$$

$$Vw_2 = 20.94 - \frac{3}{\tan 45^\circ}$$

$$Vw_2 = 17.49 \text{ m/s}$$

$$\text{Workdone per kN of water} = \frac{1}{g} Vw_2 u_2$$

$$= \frac{17.94 \times 20.94}{9.81}$$

$$\text{Workdone per kN of water} = 38.29 \text{ m}$$

Calculating velocity and direction of water at outlet from outlet velocity triangle:

$$V_2 = \sqrt{Vw_2^2 \times Vf_2^2}$$

$$V_2 = \sqrt{17.94^2 + 3^2}$$

$$V_2 = 18.19 \text{ m/s}$$

$$\tan \beta = \frac{Vf_2}{Vw_2}$$

$$\tan \beta = \frac{3}{17.94}$$

$$\beta = 9.49^\circ$$

2. The impeller of a centrifugal pump having external and internal diameters 500 mm and 250 mm respectively with an outlet 50 mm and running at 1200 r.p.m. works against a head of 48 m. The velocity of flow through the impeller is constant and equal to 3.0 m/s. The vanes are set back at an angle of 45° at outlet. Let us find Inlet Vane angle and Work-done by the impeller and Manometric efficiency.

Solution:

Given:

- $Z_2 = 500 \text{ mm} = 0.5 \text{ m}$
- $D_1 = 250 \text{ mm} = 0.25 \text{ m}$
- $B = 50 \text{ mm} = 0.05 \text{ m}$
- $N = 1200 \text{ rpm}$
- $H = 48 \text{ m}$
- $V_1 = V_{f1} = Vf_2 - 3 \text{ m/s}$
- $\varphi = 40^\circ$

$$u_1 = \frac{\pi D_1 N}{60} = \pi \times 0.25 \times \frac{1200}{60}$$

$$= 15.7 \text{ m/s}$$

$$\tan\theta = \frac{V_1}{u_1} = \frac{3}{15.7}$$

$$\theta = 10^\circ 49$$

$$u_2 = \frac{\pi D_2 N}{60} = \frac{\pi \times 0.5 \times 1200}{60}$$

$$= 31.42 \text{ m/s}$$

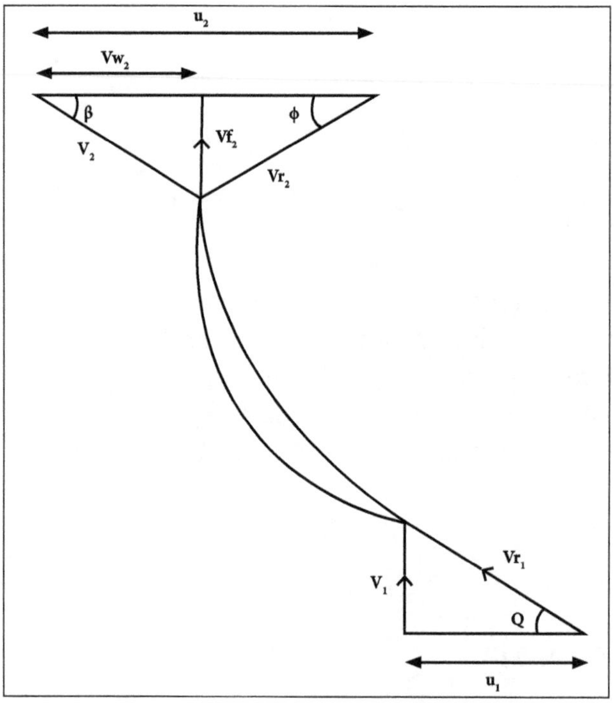

$$\tan\phi = \frac{Vf_2}{u_2 - V_{w_2}}$$

$$\tan 40^\circ = \frac{3}{31.42 - Vw_2} \Rightarrow V_{w_2} = 27.84 \text{ m/s}$$

$$\tan\beta = \frac{Vf_2}{V_{w_2}} = \frac{3}{27.84}$$

$$\beta = 6^\circ 9$$

$$\text{Workdone} = \frac{V_{w_2} u_2}{g} = \frac{27.84 \times 31.42}{9.81} = 89.17\,\text{Nm}$$

$$\eta_{man} = \frac{g H_m}{V_{w_2} u_2} \times 100 = \frac{9.81 \times 48}{27.84 \times 31.42} \times 100 = 53.83\%$$

5.2.2 Specific Speed in Centrifugal Pump

The specific speed of a centrifugal pump is defined as the speed of a geometrically similar pump which will deliver unit quantity.

$$\text{Discharge} = \text{Area} \times \text{Velocity}$$

$$Q = \pi B_1 D_1 \times vf_1$$

$$Q \alpha D_1^2 vf_1 \qquad \dots (1)$$

$$Q \alpha B_1 D_1 vf_1$$

$$(\because B \,\alpha D)$$

Also,

$$vf_1 = Kf\sqrt{2gH_m}\,(\text{or})\, vf_1 \,\alpha \sqrt{H_m} \qquad \dots (2)$$

Substitute this in equation (1) then:

$$Q \alpha D_1^2 \sqrt{H_m}$$

$$u_1 = \frac{\pi D_1 N}{60}; \Rightarrow D_1 = \frac{60 \cdot a_1}{\pi \cdot N} \qquad \dots (3)$$

$$D_1 \alpha \frac{u_1}{N}$$

$$u_1 = K_u \sqrt{2gH_m} \qquad \dots (4)$$

Substitute this in equation (4) and then we get:

$$u_1 \alpha \sqrt{H_m} \Rightarrow D_1 \alpha \frac{\sqrt{H_m}}{N} \qquad \dots (5)$$

Substitute this value in equation (3) and we get:

$$Q \alpha \left[\frac{\sqrt{H_m}}{N}\right]^2 \cdot \sqrt{H_m}$$

$$Q = \frac{H_m^{3/2}}{N^2} \Rightarrow Q = K \cdot \frac{H_m^{3/2}}{N^2} \qquad \dots (6)$$

According to definition Q = 1, Hm = 1, N = Ns then:

$$Q = K \cdot \frac{H_m^{3/2}}{N^2}$$

$$1 = K \cdot \frac{1^{3/2}}{N_s^2}$$

$$K = N_S^2 \qquad ...(7)$$

Substitute in equation (6) we get:

$$Q = K \frac{H_m^{3/2}}{N^2}$$

$$Q = N_s^2 \frac{H_m^{3/2}}{N^2}$$

$$N_s^2 = \frac{N^2 Q}{H_m^{3/2}}$$

$$N_S = \frac{N\sqrt{Q}}{H_m^{3/4}}$$

5.3 Losses and Efficiencies: Pumps in Series and Parallel

A fluid flow process involving a pump the overall efficiency is related to:

- Hydraulic efficiency.
- Mechanical efficiency.
- Volumetric efficiency.

Losses in the Pump

Hydraulic Loss and Hydraulic Efficiency

Hydraulic losses relates to the construction of the pump:

- Friction between the fluid and the walls.
- Acceleration and retardation of the fluid.
- Change of the fluid flow direction.

The hydraulic efficiency can be expressed as:

$$\eta_h = w/(w + w_I)$$

Where,

- η_h = Hydraulic efficiency.
- w = Specific work from the pump or fan (J/kg).
- w_I = Specific work lost due to hydraulic effects (J/kg).

Mechanical Loss and Mechanical Efficiency

Mechanical components like transmission gear and bearings creates mechanical losses that reduces the power transferred from the motor shaft to the pump or fan impeller.

The mechanical efficiency can be expressed as:

$$\eta_m = (P - P_l)/P$$

Where,

- η_m = Mechanical efficiency.
- P = Power transferred from the motor to the shaft (W).
- P_l = Power lost in the transmission (W).

Volumetric Loss and Volumetric Efficiency

Due to leakage of fluid between the back surface of the impeller hub plate and the casing or through other pump components, there is a volumetric loss reducing the pump efficiency.

The volumetric efficiency can be expressed as:

$$\eta_v = q/(q + q_l)$$

Where,

- η_v = Volumetric efficiency.
- q = Volume flow out of the pump or fan (m3/s).
- q_l = Leakage volume flow (m3/s).

Total Loss and Overall Efficiency

The overall efficiency is the ratio of power actually gained by the fluid to power supplied to the shaft. The overall efficiency can be expressed as:

$$\eta = \eta_h \eta_m \eta_v$$

Where, η = Overall efficiency.

The losses in a pump or fan converts to heat that is transferred to the fluid and the surroundings. As a rule of thumb, the temperature increase in a fan transporting air is approximately 1oC.

Pumps in Series and Parallel

The head or flow rate of a single pump is not sufficient for an application; pumps are combined in series or in parallel to meet the desired requirements. Pumps are combined in series to obtain an increase in head or in parallel for an increase in flow rate.

The combined pumps need not be of the same design. Figures (a) and (b) show the combined H-Q characteristic for the cases of identical pumps connected in series and parallel respectively.

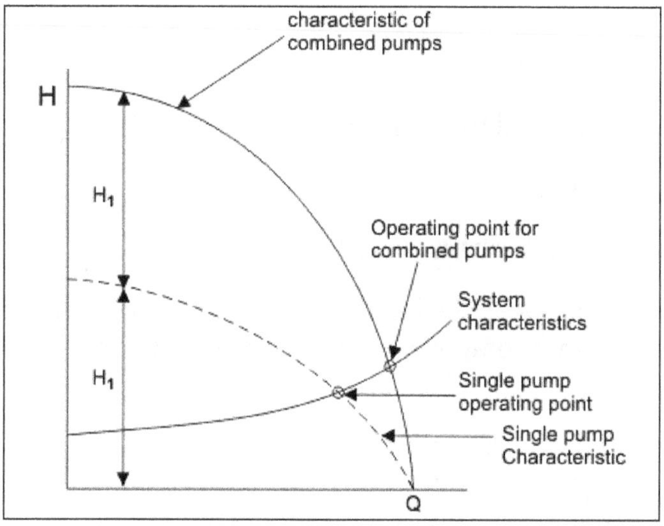

(a) Two similar pumps connected in series.

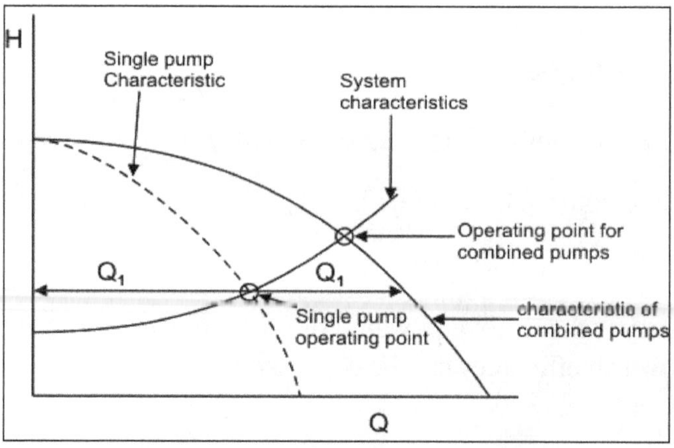

(b) Two similar pumps connected in parallel.

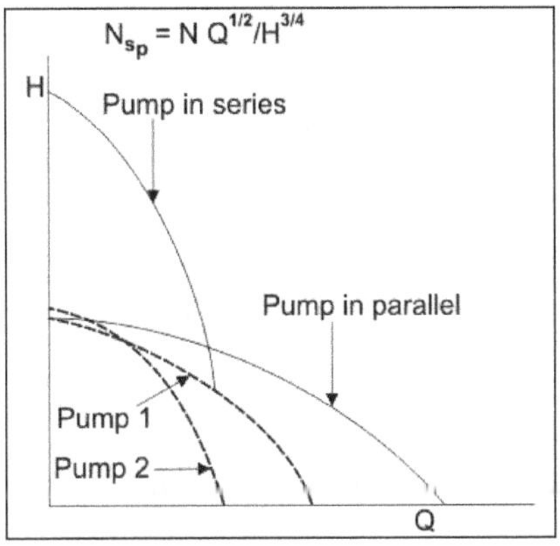

Two different pumps connected in series and parallel.

Problem

An inline water pump works between pressure 1 bar (1 105 N/m²) and 10 bar (10 105 N/m²). The density of water is 1000 kg/m³. The hydraulic efficiency is ηh = 0.91.

The actual water head (water column) can be calculated as:

$$h = (p_2 - p_1)/\gamma$$

$$= h(p_2 - p_1)/\rho g$$

$$= (p_2 - p_1)/\rho g$$

$$= ((10\ 105\ \text{N/m}^2) - (1\ 105\ \text{N/m}^2))/(1{,}000\ \text{kg/m}^3)(9.81\ \text{m/s}^2)$$

$$= 91.7\ \text{m} - \text{water column}$$

The pump must be constructed for the specific work:

$$w_c = g_h \eta_h$$

$$= (9.81\ \text{m/s}^2)(91.7\ \text{m})/0.91$$

$$= 988.6\ (\text{J/kg},\ \text{m}^2/\text{s}^2)$$

The construction or design head is:

$$h = w_c/g$$

$$= (988.6\ \text{m}^2/\text{s}^2)/(9.81\ \text{m/s}^2)$$

$$= 100.8\ \text{m} - \text{water column}$$

5.4 Performance Characteristic Curves, Cavitation and NPSH

Characteristic versus of centrifugal pumps are defined those curves which are plotted from the results of a number of tests on centrifugal pump. These curves are necessary to predict the behavior and performance of the pump when the pump is working under different flow rate, head and speed.

The followings are important characteristic curves for pumps:

- Main characteristic curves.
- Operating characteristic curves.
- Constant efficiency curve.
- Constant head and constant discharge curves.

Main Characteristic Curve

The main characteristic curves of a centrifugal pump consist of variation of Manometric head (m), power and discharge with respect to speed. For plotting curves of manometric head versus speed, discharge is kept constant. For plotting curves of discharge versus speed, manometric head (m) is kept constant. And for plotting curves of power versus speed, the manometric head and discharge are kept constant.

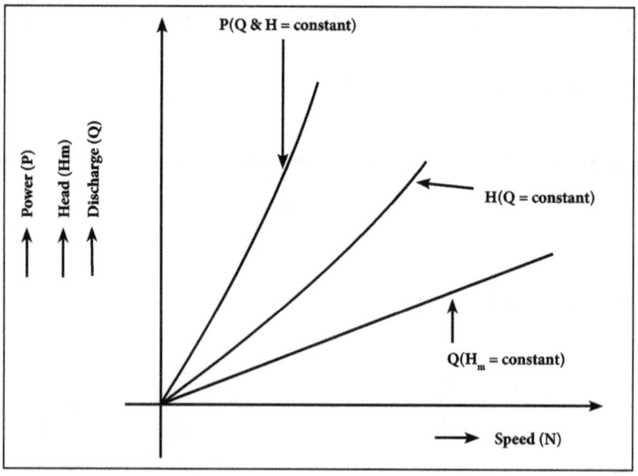

Characteristic curves of a centrifugal pump.

Operating Characteristic Curve

If the speed is kept constant, the variation of manometric head, power and efficiency with respect to discharge gives the operating characteristics of the pump.

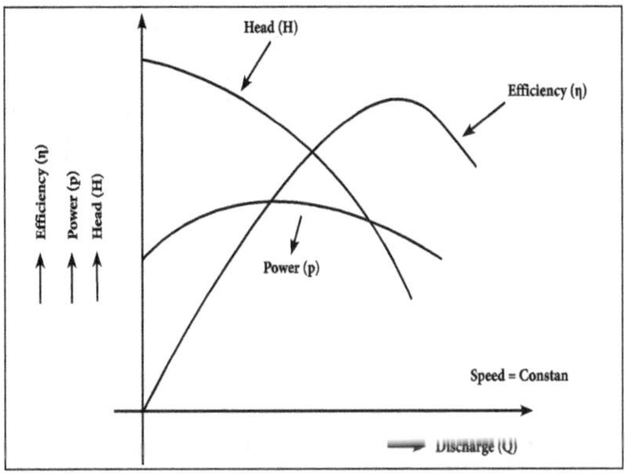

Operating characteristics of the pump.

Constant Efficiency Curve

For obtaining constant efficiency curves for a pump-the head versus discharge curves and efficiency versus discharge curves for different speeds are used. Figure shows the head versus discharge curves for different speeds. The efficiency versus discharge curves for the different speeds are shown in figure. By combining these curves (H ≈ Q curves and η ≈ Q curves), constant efficiency curves are obtained as shown in figure:

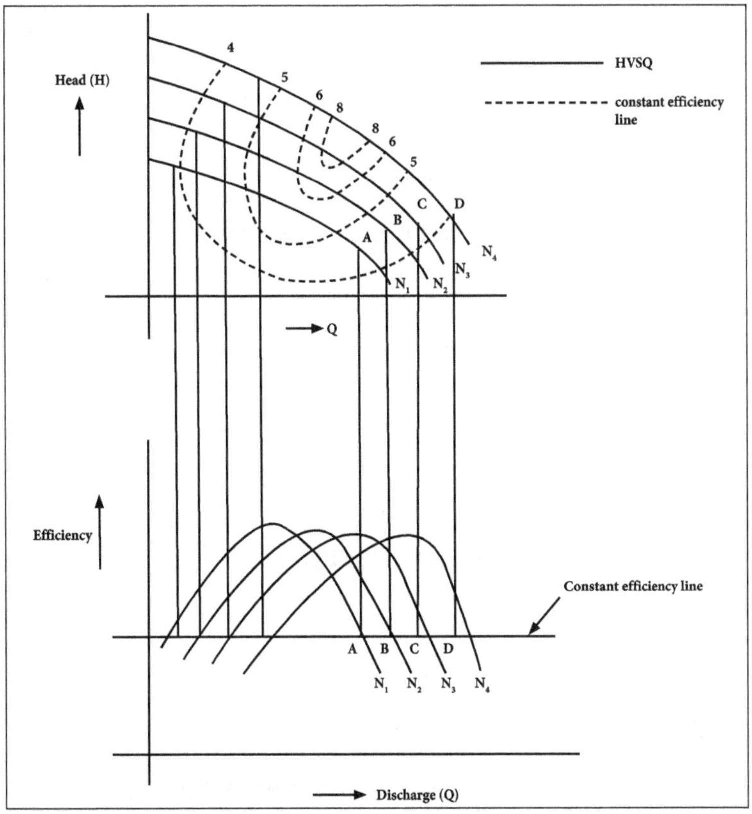

Constant efficiency curves for a pump.

Constant Head and Constant Discharge Curves

The performance of a variable speed pump is for which the constant speed variation can be obtained by these curves. When the pump has a variable speed, the plotted graph between Q and N and Hm and N may be obtained. In the first case, Hm is kept constant and in second case Q is kept constant.

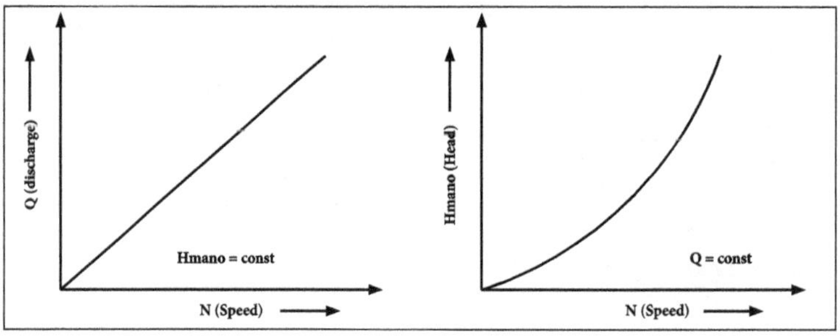

Constant head and constant discharge curves.

Cavitation's in Pumps

Cavitation occurs when a liquid passing through a pump vaporizes. Cavitation occurs when there is low pressure or high suction. Cavitation reduces the efficiency of the pump and can increase wear and tear on the pump. "Chemical Engineering Fluid Mechanics" by Ron Darby states, "to prevent cavitation, it is necessary that the pressure at the pump suction be sufficiently high that the minimum pressure anywhere in the pump will be above vapor pressure." Preventing cavitation can be done by increasing decreasing pressure at the intake or increasing pressure at the discharge as well as reducing the sources of potential bubbles.

Cavitation is a problem condition which may develop while a centrifugal pump is operating. This occurs when a liquid boils inside the pump due to insufficient suction head pressure. Low suction head causes a pressure below that of vaporization of the liquid, at the eye of the impeller. The resultant gas which forms causes the formation and collapse of 'bubbles' within the liquid. because gases cannot be pumped together with the liquid, causes violent fluctuations of pressure within the pump casing and is seen on the discharge gauge.

These sudden changes in pressure cause vibrations which can result in serious damage to the pump and of course cause pumping inefficiency.

To overcome cavitation:

- Increase suction pressure if possible.
- Decrease liquid temperature if possible.
- Throttle back on the discharge valve to decrease flow-rate.
- Vent gases off the pump casing.

5.5 Reciprocating Pumps

Displacement of Reciprocating Pump

The reciprocating pump is called a positive displacement pump because the liquid is sucked and then displaced in a piston-cylinder arrangement for lifting liquids to required height. It is driven by natural source.

5.5.1 Working of Reciprocating Pump

According to the water being in content with Piston this Pump classified into two types:

- Single Acting reciprocating Pump.
- Double Acting reciprocating Pump.

Working Procedure of Single Acting Reciprocating Pump

It has one suction pipe and delivery pipe; it is usually placed above the liquid level in the Sump. When the crank rotates clockwise from inner dead center (IDC) to out of dead center [ODC]. The piston moves outwards to the right and a vacuum is created on the left side of the piston.

This vacuum causes suction value to open and liquid is forced. From the sump into the left side of the Piston. When the crank is at ODC, the suction stroke is completed and the left side of cylinder is full of liquid. When the crank rotates from ODC to IDC. The Piston moves inward to the left and a high Pressure is built up in the cylinder.

The increased pressure causes the suction valve to close and the discharge valve to open. Thus the liquid is carried to the discharge tank through the delivery Pipe. The delivery stroke is completed when the crank occupies the ODC Position. The suction and delivery stroke alternatively carried out and the liquids pumped from the sump to the discharge tank.

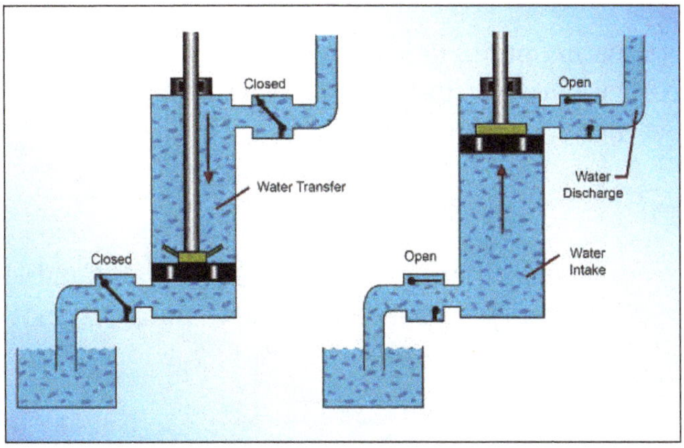

Single Acting Reciprocating Pump.

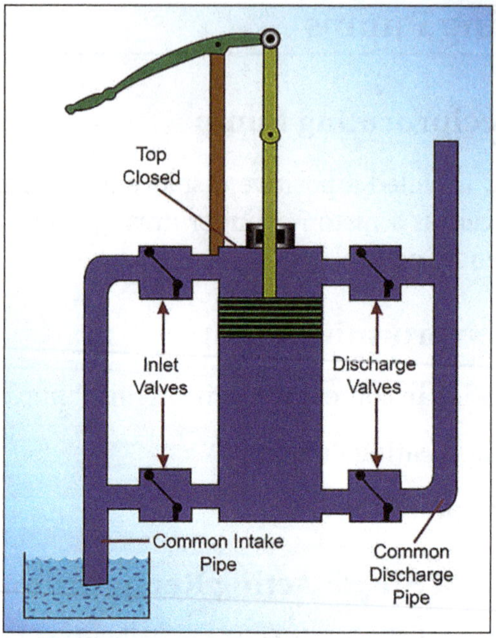

Double acting Reciprocating pump.

Working Procedure of Double Acting Reciprocating Pump

The double acting reciprocating pump, the suction and delivery stroke occurs simultaneously when the crank rotates from IDC to ODC, a vacuum is created to the left side of the piston and the liquid is sucked from the sump through suction value S. At the same time, the liquid on the right side of the piston is pressed and a high pressure causes the delivery value D2 it open and the liquid is passed to the discharge tank. This operation container all the crank reaches to ODC.

With further rotation of the crank, the liquid is sucked from the sump the sump through suction value S2 and delivered to the different tank through delivery value D, When the crank reaches IDC, the is in the extreme left position. Thus one cycle is completed.

Double acting reciprocating pump quick move uniform discharge because of continuous delivery. Quite after a multi cylinder arrangements having two are move cylinder is also employed to get more.

5.5.2 Discharge

The capacity of a reciprocating pump is the cylinder displacement less slip. For a single-acting cylinder, cylinder displacement can be determined from:

$$s = \frac{(A \times L_s \times N \times m)}{231} \qquad ...(1)$$

For double-acting cylinders, the cylinder displacement can be determined by:

$$s = \frac{\left[(2A - a) \times L_s \times N \times m\right]}{231} \qquad ...(2)$$

Where,

- s = Cylinder displacement.
- A = Plunger or piston area.
- a = Piston-rod cross-sectional area.
- L_S = Stroke length.
- N = Speed.
- m = Number of pistons or plungers.

The pump capacity can be determined from:

$$q = \frac{s(100 - S)}{100} \qquad ...(3)$$

Where, q = Pump capacity.

5.5.3 Indicator Diagram for Reciprocating Pump

The indicator diagram for a reciprocating pump is defined as the graph between the pressure fixed head in the cylinder and the distance traveled by piston from inner dread center for one complete revolution of the crank. As the maximum distance traveled by the piston is equal to the stroke length and hence the indicator diagram is a graph between pressure head and stroke length of the piston for one complete revolution.

The graph between pressure head in the cylinder and stroke length of the piston for one complete revolution of the crank under ideal conditions is known as ideal indicator which line EF represents the atmospheric pressure head equal to 10.3 m of water.

Let,

- $H_{atm} \rightarrow$ Atmospheric pressure head = 10.3 m of water.
- $L \rightarrow$ Length of the stroke.
- $h_s \rightarrow$ Suction head.
- $h_d \rightarrow$ Delivery head.

During suction stroke, the pressure head in the cylinder is constant and equal to suction head (ls), which is below atmospheric pressure head (Ha tm) by a height of hs. The pressure head during suction stroke is represented by a horizontal line AB which is below the line EF by a height of 'hs'.

During delivery stroke, the pressure head in the cylinder is constant and equal to delivery head (hd), which is above the atmospheric head by a height of (hd). Thus the pressure head during delivery stroke is represented by a horizontal line CD which is above the line EF by a height of hd. Thus, for one complete revolution of crank, the pressure head in the cylinder is represented by the diagram A–B–C–D–A. This diagram is known as ideal indicator diagram.

We know that work done by pump per second:

$$\rho = \frac{g \times ALN}{60} \times (h_s + h_d)$$

$$= k \times L(h_s + h_d) \left(\because \frac{p_g AN}{60} \right)$$

$$= \alpha L \times (h_s + h_d) \qquad \text{...(1)}$$

But,

Area of indicator diagram $= AB \times BC$

$$= AB \times (BF + FC)$$

$$= L \times (h_s + h_d) \qquad \text{... (2)}$$

Work done by pump α Area of indicator diagram:

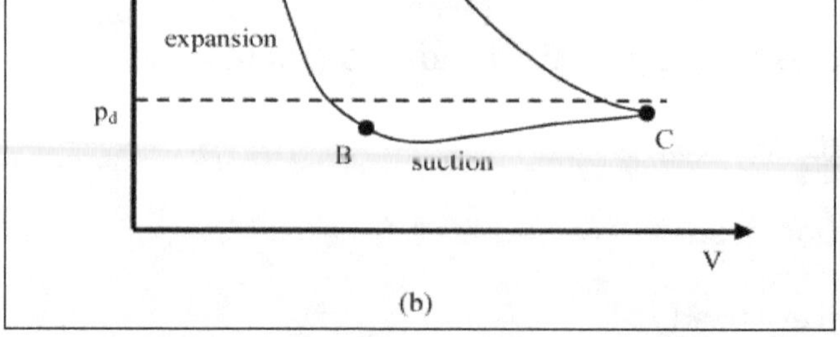

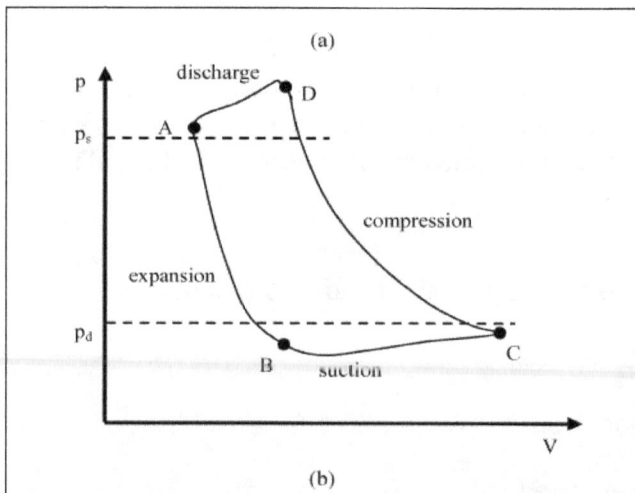

Schematic of a reciprocating compressor and its indicator diagram.

5.5.4 Slip of Reciprocating Pump

Slip of a pump is defined as the difference between the theoretical discharge and actual discharge of the pump.

$$\text{Slip} = Q_{th} - Q_{act}$$

Percentage slip:

$$\% \text{ of Slip} = \frac{Q_{th} - Q_{act}}{Q_{th}} = \left(1 - \frac{Q_{act}}{Q_{th}}\right) = 100$$

$$= \left(1 - C_d \times 100\right)$$

$$CB_d = \text{Co} - \text{efficient of discharge}$$

$$= \frac{Q_{act}}{Q_{th}}$$

Negative Slip:

If the actual discharge is more than the theoretical discharge, the slip of the Pump will become –ive. In that case, the slip of the Pump of known as negative slip. Negative slip occurs when delivery Pipe is short, suction Pipe is long and pump is running at high speed.

Problems

1. The diameter and stroke of a single acting reciprocating pump are 120 mm and 300 mm respectively. The water is lifted by a pump through a total head of 25 m. The diameter and length of delivery pipe are 100 mm and 20 m respectively. Let us find out (i) Theoretical discharge and theoretical power required to run the pump if its speed is 60 rpm (ii) Percentage slip, if the actual discharge is 2.95 I/s and (iii) The acceleration head at the beginning and middle of the delivery stroke.

Solution:

Given data:

$$D = 120 \text{ mm} = 0.12 \text{ m}$$

$$L = 300 \text{ mm} = 0.3 \text{ m}$$

$$\therefore r = \frac{L}{2} = \frac{0.3}{2} = 0.15 \text{ m}$$

$$N = 60 \text{ rpm}$$

$$Q_{actual} = 2.95 \text{ liters/s} = 0.00295 \text{ m}^3/\text{s}$$

$$h_d = 25 \text{ m}$$

$$l_d = 20 \text{ m}$$

$$d_d = 100 \text{ mm} = 0.10 \text{ m}$$

To Find:

- Theoretical discharge.
- Percentage slip.
- Acceleration head at the beginning of the stroke.

Formula to be used:

$$Q_{th} = \frac{ALN}{60}$$

$$= \frac{\frac{\pi}{4} \times 0.12^2 \times 0.3 \times 60}{60}$$

$$\% \text{ slip} = Q_{th} - \frac{Q_{act}}{Q_{th}} \times 100$$

$$= \frac{0.00339 - 0.00295}{0.00339} \times 100$$

$$\% \text{ slip} = 12.98\%$$

$$h_{ad} = \frac{l_d}{g} \frac{A}{a_d} \omega^2 r \qquad \begin{bmatrix} \because \quad \theta = 0^\circ \\ \text{Cos}\, 0 = 1 \end{bmatrix}$$

$$= \frac{20}{9.81} \times \frac{\frac{\pi}{4}(0.12)^2}{\frac{\pi}{4}(0.1)^2} \times \left(\frac{2\pi \times 60}{60}\right)^2 \times 0.15$$

$$= 17.38 \text{ m of water}$$

(i) Theoretical discharge:

$$Q_{th} = \frac{ALN}{60}$$

$$= \frac{\frac{\pi}{4} \times 0.12^2 \times 0.3 \times 60}{60}$$

$$= 0.00339 \text{ m}^3/\text{s}$$

Power required to run the pump:

$$P = w\,Q_{th}\left(h_s + h_d\right)$$

$$= 9810 \times 0.00339(0+25)$$

$$P = 0.8313\ \text{kW}$$

(ii) Percentage slip:

$$\%\ \text{slip} = Q_{th} - \frac{Q_{act}}{Q_{th}} \times 100$$

$$= \frac{0.00339 - 0.00295}{0.00339} \times 100$$

$$\%\ \text{slip} = 12.98\%$$

(iii) Acceleration head at the beginning of the stroke:

$$h_{ad} = \frac{l_d}{g}\frac{A}{a_d}\omega^2 r \qquad \begin{bmatrix} \because & \theta = 0^\circ \\ \text{Cos}\,0 = 1 \end{bmatrix}$$

$$= \frac{20}{9.81} \times \frac{\frac{\pi}{4}(0.12)^2}{\frac{\pi}{4}(0.1)^2} \times \left(\frac{2\pi \times 60}{60}\right)^2 \times 0.15$$

$$= 17.38\,\text{m of water}$$

Acceleration head at the middle of the stroke:

$$h_{ad} = \frac{l_d}{g}\frac{A}{a_r}\omega^2 r \cos 90^\circ$$

$$h_{ad} = 0$$

2. A single acting reciprocatory pump having a plunger of 30 cm diameter and a stroke of 20 cm. If the speed of the pumps is 30 rpm and it delivers to 6.5 lit/s of water, let us determine the coefficient of discharge and the percentage slip of the pump.

Solution:

Given:

- Piston diameter, $D = 30\ \text{cm} = 0.3\ \text{m}$
- Stroke, $L = 20\ \text{cm} = 0.2\ \text{m}$

- Speed, $N = 20$ rpm
- Actual Discharge, $Q_a = 6.5$ lit/s $= 6.5 \times 10^{-3}$ m^3/s

To find:

- Co-efficient of discharge (Cd).
- Percentage slip of pump (S).

Formula to be used:

$$C_d = \frac{\text{Actual discharge}(Q_a)}{\text{Theoretical discharge}(Q_t)}$$

$$Q_a = 6.5 \times 10^{-3} \text{ m}^3/\text{s}$$

$$Q_2 = \frac{\pi}{4} \times D^2 \times L \times \frac{N}{60}$$

$$= \frac{\pi}{4} \times 0.3^2 \times 0.2 \times \frac{30}{60}$$

$$\therefore Q_t = 7.06 \times 10^{-3} \text{ m}^3/\text{s}$$

$$\therefore C_d = Q_a / Q_t$$

$$= \frac{6.5 \times 10^{-3}}{7.6 \times 10^{-3}}$$

$$\% \text{ slip} = \frac{Q_t - Q_a}{Q_t} \times 100$$

$$= \frac{(7.6 \times 10^{-3}) - (6.5 \times 10^{-3})}{7.6 \times 10^{-3}} \times 100$$

Co-efficient of discharge, $C_d = 0.92$

$$\% \text{ slip} = \frac{Q_t - Q_a}{Q_t} \times 100$$

$$= \frac{(7.6 \times 10^{-3}) - (6.5 \times 10^{-3})}{7.06 \times 10^{-3}} \times 100$$

$$= 0.07932 \times 100$$

$$= 7.93\%$$

$$\% \text{ slip} = 7.93\%$$

Result:

- Co-efficient of discharge $C_d = 0.92$
- $\%\ \text{slip} = 7.93\%$

3. A single acting reciprocating pump running at 50 rpm, delivers 0.01 m3/s of water. The diameter of the piston is 200 mm and stroke length 400 mm. Let us determine the theoretical discharge of the pump, coefficient of discharge and slip and the percentage slip of the pump.

Solution:

Given:

- Speed of the Pump, N = 50 rpm.
- Actual discharge Q act = 0.01m3/s.
- Dia. of piston D = 200 mm = 0.20 m.

To find:

- Theoretical discharge for single-acting reciprocating pump.
- Coefficient of discharge.

Formula to be used:

$$Q_{th} = \frac{A \times L \times N}{60} = \frac{0.031416 \times 0.40 \times 50}{60} = 0.01047\ m^3/s$$

$$C_d = \frac{\phi_{act}}{\phi_{th}} = \frac{0.01}{0.01047} = 0.955$$

$$\therefore \text{Area}(A) = \frac{\pi}{4} \cdot (0.2)^2 = 0.031416\, m^2$$

Storage, $L = 400\ \text{mm} = 0.40\ \text{m}$

(i) Theoretical discharge for single-acting reciprocating pump is given by equation as:

$$Q_{th} = \frac{A \times L \times N}{60} = \frac{0.031416 \times 0.40 \times 50}{60} = 0.01047\ m^3/s$$

(ii) Coefficient of discharge is given by:

$$C_d = \frac{\phi_{act}}{\phi_{th}} = \frac{0.01}{0.01047} = 0.955$$

(iii) Using equation, we get:

$$\text{Slip} = \phi_{th} - \phi_{act} = 0.01047 - 0.01 = 0.00047 \text{ m}^3/\text{s}$$

(iv) Percentage slip:

$$\% \text{ Slip} = \frac{(\phi_{th} - \phi_{act})}{\phi_{th}} \times 100 = \frac{(0.01047 - 0.01)}{0.01047} \times 100$$

$$= \frac{0.00047}{0.01047} \times 100 = 4.487\%$$

6

Hydraulic Turbines

6.1 Classification of Turbines: Impulse and Reaction Turbines

Classification of Turbines

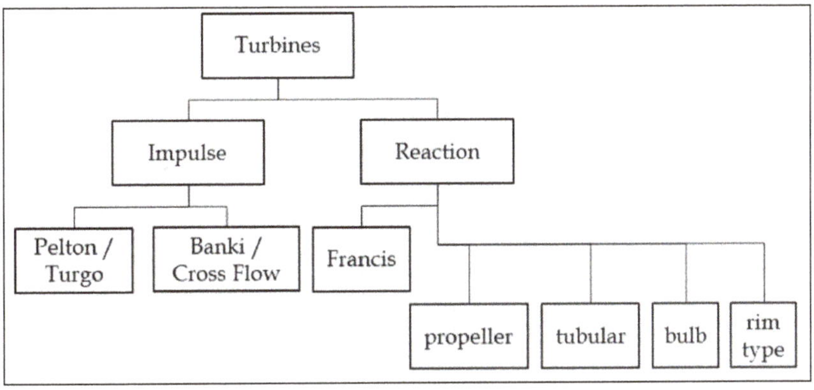

There are two main types of hydro turbines: Impulse and reaction.

The type of hydro power turbine selected for a project is based on the height of standing water referred to as "head" and the flow or volume of water, at the site. Other deciding factors include how deep the turbine must be set, efficiency and cost.

Impulse Turbine

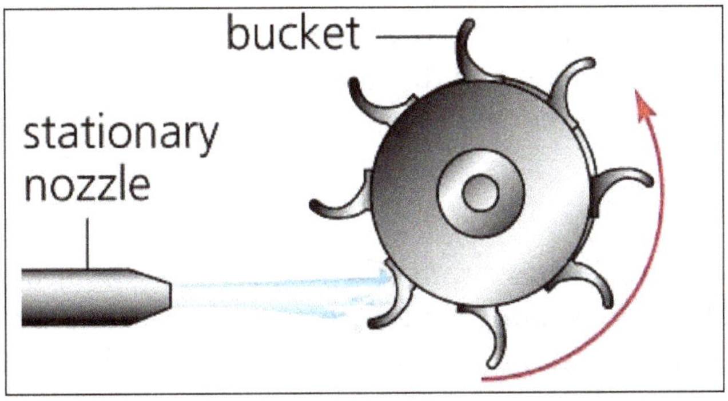

Impulse Turbine.

The impulse turbine generally uses the velocity of the water to move the runner and discharges to atmospheric pressure. The water stream hits each bucket on the runner. There is no suction on the down side of the turbine and the water flows out the bottom of the turbine housing after hitting the runner. An impulse turbine is generally suitable for high head and low flow applications.

Reaction Turbine

In a reaction turbine, the blades sit in a much larger volume of fluid and turn around as the fluid flows past them. A reaction turbine doesn't change the direction of the fluid flow as drastically as an impulse turbine: it simply spins as the fluid pushes through and past its blades. Wind turbines are perhaps the most familiar examples of reaction turbines.

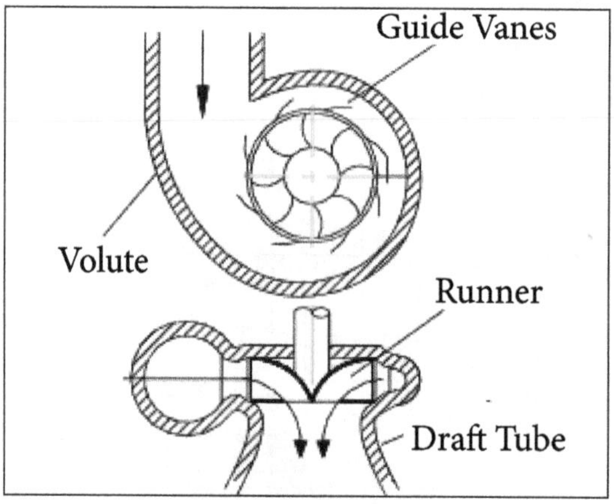

Reaction Turbine.

The principal feature of a reaction turbine that distinguishes it from an impulse turbine is that only a part of the total head available at the inlet to the turbine is converted to velocity head, before the runner is reached. Also in the reaction turbines the working fluid, instead of engaging only one or two blades, completely fills the passages in the runner.

The pressure or static head of the fluid changes gradually as it passes through the runner along with the change in its kinetic energy based on absolute velocity due to the impulse action between the fluid and the runner. Therefore the cross-sectional area of flow through the passages of the fluid.

Propeller

A propeller turbine generally has a runner with three to six blades in which the water contacts all of the blades constantly. Picture a boat propeller running in a pipe. Through the pipe, the pressure is constant, if it isn't, the runner would be out of balance. The pitch of the blades may be fixed or adjustable. The major components besides the runner are a scroll case, wicket gates and a draft tube.

There are several different types of propeller turbines:

- Bulb turbine: The turbine and generator is a sealed unit placed directly in the water stream.
- Tube turbine: The pen-stock bends just before or after the runner, allowing a straight line connection to the generator.
- Kaplan: Both the blades and the wicket gates are adjustable, allowing for a wider range of operation.
- Francis: A Francis turbine has a runner with fixed buckets (vanes), usually nine or more. Water is introduced just above the runner and all around it and then falls through, causing it to spin. Besides the runner, the other major components are the scroll case, wicket gates and draft tube.
- Kinetic: Kinetic energy turbines, also called free-flow turbines, generate electricity from the kinetic energy present in flowing water rather than the potential energy from the head. The systems may operate in rivers, manmade channels, tidal waters or ocean currents. Kinetic systems utilize the water stream's natural pathway.

They do not require the diversion of water through man made channels, riverbeds or pipes, although they might have applications in such conduits. Kinetic systems do not require large civil works, however, they can use existing structures such as bridges, tail races and channels.

Table: Comparison between turbines and pumps.

S. No.	Pumps	Turbines
1.	In pumps flow takes place from the low pressure to high pressure.	In turbines flow takes place from high pressure to low pressure.
2.	Pump flow is accelerated flow.	In turbine, there is a decelerated flow.
3.	It is energy absorbing machine.	It is energy producing machine.

The hydraulic turbines are classified as follows:

- According to the head quantity of water available.
- According to the name of the originator.
- According to the action of water on moving blades.
- According to the direction of flow of water in the runner.

- According to the disposition of the turbine shaft.
- According to the specific speed N.

Classify turbines according to flow:

- Tangential flow turbine. Ex. Pelton wheel turbine.
- Radial flow turbine. Ex. Old Francis turbine.
- Axial flow turbine. Ex. Kaplan turbine.
- Mixed flow turbine. Ex. Modem Francis turbine.

Table: Differentiation of impulse and reaction turbine.

S. No.	Impulse Turbine	Reaction Turbine
1.	All the potential energy is converted into kinetic energy by nozzle before entering to turbine runner.	Only a portion of the fluid energy is transferred into kinetic energy before the fluid enters the turbine.
2.	Flow regulation is possible without loss.	Flow regulation is possible with loss.
3.	Flow is regulated by means of a needle valve fitted into the nozzle.	Flow is regulated by means of a guide-vane assembly.
4.	Water may be allowed to enter a part or whole of the wheel circumference.	Water is admitted over the circumference of the wheel.
5.	Wheel does not run full and air has free access to the buckets.	Water completely fills the vane passages throughout the operation of the turbine.
6.	Unit is installed above the tail race.	Unit is kept entirely submerged in water below tail race.
7.	Blades are only in action when they are in front of nozzle.	Blades are in action at all the time.

6.1.1 Heads and Efficiencies

Volumetric and mechanical efficiencies:

Volumetric Efficiency:

$$\eta_v = \frac{\text{Volume of water actually striking the runner}}{\text{Volume of water supplied the turbine}} = \frac{Q_a}{Q}$$

Mechanical Efficiency:

$$\eta_m = \frac{\text{Power at the shaft of the turbine}}{\text{Power delivered by water the runner}} = \frac{P}{wQ_a \cdot H_r}$$

$$\eta_m = \frac{S_P}{R.P}$$

Hydraulic Efficiency:

$$\eta_h = \frac{\text{Power delivered runner}}{\text{Power supplied at inlet}} = \frac{H_r}{H}$$

$$\eta_h = \frac{R.P}{W.P}$$

Overall Efficiency:

$$\eta_o = \frac{\text{Power available at the turbine shaft}}{\text{Power available the water jet}}$$

$$= \frac{\text{Shaft power}}{\text{Water power}} = \frac{P}{wQH}$$

$$= \frac{P}{wQH}$$

Problems

1. In a hydroelectric station, water is available at the rate of 175 m3/s under head of 18 m. The turbine run at a speed of 150 rpm, with overall efficiency of 82%. Let us determine the number of turbines required, if they have the maximum specific speed of 460.

Solution:

Given:

$$Q = 175\,m^3/s$$

$$H = 18\,m$$

$$N = 150 \text{ rpm}$$

$$\eta_o = 82\%$$

$$N_s = 460$$

To Find: The Number of Turbines Required.

Formula to Be Used:

$$N_s = \frac{N\sqrt{P}}{H5/4}$$

Number of turbines required: Specific speed of the turbine:

$$N_s = \frac{N\sqrt{P}}{H5/4}$$

$$460 = \frac{150\sqrt{P}}{(18)5/4}$$

Power available at turbine shaft:

$$P = \left[\frac{460 \times (18)^{5/4}}{150}\right]^2$$

$$= 12927.5\,\text{kW}$$

Power available from turbines $= WQH \times \eta_o$

$$= 9.81 \times 175 \times 18 \times 0.82$$

$$= 25339.23\ \text{kW}$$

No. of turbines required$= \dfrac{25339.23}{12927.5}$

$$= 1.96 \approx 2\ \text{Nos.}$$

2. A Pelton wheel having a mean bucket diameter of 1 m runs at 1000 rpm. The non-head on the Pelton wheel is 700 m. If the side clearance angle is 15° and discharge through nozzle is 0.1 m3/s, let us determine (1) power available at the nozzle and (2) hydraulic efficiency of the turbine. Take C = 1.

Solution:

Given:

Diameter of wheel, $D = 1.0\,\text{m}$

Speed of wheel, $N = 1000$ r.p.m

To find:

Power available at the nozzle.

Hydraulic efficiency of the turbine.

Formula to be used:

$$W.P = \frac{W \times H}{1000} = \frac{\rho \times g \times Q \times H}{1000}$$

$$= \frac{1000 \times 9.81 \times 0.1 \times 700}{1000} = 686.7\,KW$$

$$\eta h = \frac{2(V_1 - u)(1 + \cos\phi)u}{V_1^2}$$

$$= \frac{2(117.18 - 52.36)(1 + \cos 15) \times 52.36}{117.19 \times 117.19}$$

$$= \frac{2 \times 64.83 \times 1.966 \times 52.36}{117.19 \times 117.19} = 0.9718 = 97.18\%$$

∴ Tangential velocity of the wheel:

$$u = \frac{\pi DN}{60} = \frac{\pi \times 1.0 \times 1000}{60} = 52.36\,m$$

Net head or turbine $H = 700$ m

Side clearance angle $\phi = 15^\circ$

Discharge, $Q = 0.1\ m^3/s$

Velocity of jet at inlet:

$$V_1 = C_v\sqrt{2gH} = 1 \times \sqrt{2 \times 9.81 \times 700}$$

(∴ Value of Cv is not given. Take it = 1.0)

Or, $V_1 = 117.19\ m/s$

1. Power available at the nozzle is given by equation as:

$$W.P = \frac{W \times H}{1000} = \frac{\rho \times g \times Q \times H}{1000}$$

$$=\frac{1000\times 9.81\times 0.1\times 700}{1000}=686.7\,\text{KW}$$

2. Hydraulic efficiency is given by equation as:

$$\eta h=\frac{2(V_1-u)(1+\cos\phi)u}{V_1^2}$$

$$=\frac{2(117.18-52.36)(1+\cos 15)\times 52.36}{117.19\times 117.19}$$

$$=\frac{2\times 64.83\times 1.966\times 52.36}{117.19\times 117.19}=0.9718=97.18\%$$

6.1.2 Velocity triangles

In turbo machinery is a velocity triangle or velocity diagram is a triangle representing the various components of velocities of the working fluid in a turbo machine. Velocity triangles may be drawn for both the inlet and outlet sections of any turbo machine.

The vector nature of velocity is utilized in the triangles and the most basic form of a velocity triangle consists of the tangential velocity, the absolute velocity and the relative velocity of the fluid making up three sides of the triangle.

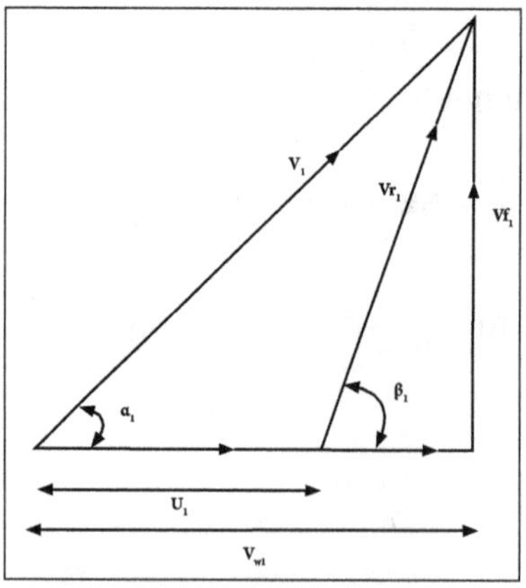

Velocity triangle of turbine.

Velocity Involved

- V - Absolute velocity of the fluid.
- U - Blade Linear velocity.
- V_r - Relative velocity of the fluid after contact with rotor.
- V_w - Tangential component of V (absolute velocity) called Whirl velocity.

- V_f - Flow velocity (axial component in case of axial machines, radial component in case of radial machines).

The following angles are encountered during the analysis:

- α = Angle made by V with the plane of the machine (usually the nozzle angle or the guide blade angle).
- β = Angle of the rotor blade.

6.2 Pelton Wheel, Francis and Kaplan Turbine

6.2.1 Pelton Wheel

A Pelton wheel turbine consists of a rotor at the periphery which is mounted on equally spaced double hemispherical or double ellipsoidal buckets. Water is transferred from a high head source through pen stock which is fitted with a nozzle through which the water flows out as a high jet.

A needle spear moving inside the nozzle controls the water flows through the nozzle at the same time. It provides a smooth flow with negligible energy loss. All the potential energy available is converted into kinetic energy before the jet strikes the buckets of the runner. The pressure all over the wheel is constant and equal to the atmosphere.

The Pelton wheel is provided with a casing. The function of casing is to prevent the splashing of water and to discharge water to the tail race.

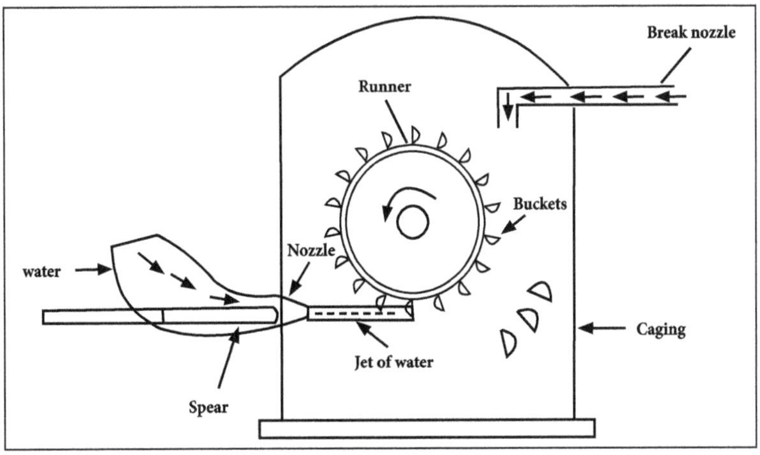

Pelton wheel.

The nozzle is completely closed by noting the spear in the forward direction of a amount of water striking the runner is reduced to zero but the runner due to inertia continues revolving for a long time. In order to bring the runner to rest in a short time, a nozzle is provided which directs the jet of water on the buckets. This jet of water is called breaking jet.

Speed of the turbine is kept constant by a governing mechanism that automatically regulates the quantity of water flowing through the runner in accordance with any variation of load.

Mechanical Efficiency

It is defined as the ratio of the power obtained from the shaft of the turbine to the power developed by the runner. These two powers differ by the amount of mechanical losses. Viz. bearing friction.

$$\eta_m = \frac{\text{Power available of the turbine shaft}}{\text{Power developed by turbine runner}}$$

$$= \frac{P}{\omega Q_a \cdot H_r}$$

A value of mechanical efficiency for a Pelton wheel usually lies between 97 to 99%. The mechanical efficiency will be more if the mechanical loss is less and capacity of the unit is high.

Problems

1. The mean velocity of the buckets of the Pelton wheel is 10 m/s. The jet supplies water at 0.7 m3/sat a head of 30 m. The jet is deflected through an angle of 160° by the bucket. Let us determine the hydraulic efficiency $C_v = 0.98$.

Solution:

Given:

Mean velocity $= 10 \text{ m/s}$

$$Q = 0.7 \text{ m}^3/\text{s}$$

Deflected angle $= 160°$

Formula to be used:

Mean velocity of buckets:

$$Q = 0.7 \text{m}^3/\text{s}$$

Angle of deflection $=160^\circ$

$$\therefore \text{Angle}, \phi = 180^\circ - 160^\circ = 20^\circ$$

Coefficient of velocity, $C_v = 0.98$

$$H = 30\,m$$

$$\text{Velocity of jet}, V_1 = C_v\sqrt{2gH}$$

$$= 0.98\sqrt{2\times 9.81\times 30}$$

$$V_1 = 23.77\,m/s$$

$$V_{\omega 1} = V_1 = 23.77\,m/s$$

$$V_{r1} = V_1 - u_1 = 23.77 - 10$$

$$V_{r1} = 13.77\,m/s$$

$$V_{r2} = V_{r1} = 13.77\,m/s$$

$$V_{\omega 2} = V_{r2}\cos\phi - u_2$$

$$= 13.77\cos 20^\circ - 10$$

$$V_{\omega 2} = 2.94\,m/s$$

The hydraulic efficiency of the turbin:

$$\eta_h = \frac{2\left[V_{w1} + V_{w2}\right]}{V_1^2}\times u$$

$$= \frac{2\left[23.77 + 2.94\right]}{23.77\times 23.77}\times 10$$

$$\eta_h = 0.9454\,(\text{or})\,94.54\%$$

$$u = u_1 = u_2 = 10\,m/s$$

2. A Pelton wheel supplies water from reservoir under a gross head of 112 m and the friction losses in the pen stock amounts to 20 m of head. The water from pen stock is discharged through a single nozzle of diameter of 100 m at the rate of 0.30 m3/s. Mechanical losses due to friction amounts to 4.3 KW of power and shaft power available is

208 KW. Let us determine: Velocity of jet, water power at inlet to runner, power loss in nozzles, power lost in runner due to hydraulic resistance.

Solution:

Given Data:

- $H_g = 112$ m
- $h_f = 20$ m
- $d = 100$ mm $= 0.1$ m
- $Q = 0.30\ m^3 / S$

To Find:

- Velocity of Jet.
- Water Power at Inlet to Runner.
- Power Loss in Nozzles.
- Power Lost In Runner Due To Hydraulic Resistance.

Formula to Be Used:

1. $V_1 = \frac{Q}{a} = \frac{0.30}{0.00785}$

$V_1 = 38.197\,m/s$

2. $W.P. = \frac{\rho g Q H}{1000}$

$= \frac{1000 \times 9.81 \times 0.30 \times 92}{1000}$

$W.P. = 270.756\,kW$

3. Power of jet + Power lost in nozzle.

4. Power at the shaft +Power lost in nozzle + Power lost in runner + Power lost due to mechanical resistance.

Power lost in Mechanical friction = 0.45 kW.

Shaft power = 208 kW.

1. Velocity of Jet:

Total head, $H = H_g - h_f$

$= 112 - 20$

$H = 92 \text{ m}$

Area of Jet, $a = \dfrac{\pi}{4} d^2$

$= \dfrac{\pi}{4} \times (0.1)^2$

$a = 0.00785 \text{ m}^2$

$Q = a \times V_1$

$V_1 = \dfrac{Q}{a} = \dfrac{0.30}{0.00785}$

$V_1 = 38.197 \text{ m/s}$

2. Water Power at inlet to runner:

$$W.P. = \frac{\rho g Q H}{1000}$$

$$= \frac{1000 \times 9.81 \times 0.30 \times 92}{1000}$$

$$W.P. = 270.756 \text{ kW}$$

Power corresponding to kinetic energy of Jet in kW:

$$= \frac{1}{2} \frac{mV^2}{1000}$$

$$= \frac{1}{2} \frac{(\rho \times a V_1^2) V_1^2}{1000}$$

$$= \frac{1}{2} \frac{\rho Q V_1^2}{1000}$$

$$= \frac{1}{2} \times \frac{1000 \times 0.30 \times (38.197)^2}{1000}$$

$$= 218.85 \text{ kW}$$

3. Power lost in Nozzle:

Power at the base of nozzle = Power of jet + Power lost in nozzle

270.756 = 218.85 + Power lost in nozzle

Power lost in Nozzle = 51.90 kW

4. Power lost in runner due to hydraulic resistance:

Power at the base of Nozzle = Power at the shaft +Power lost in nozzle + Power lost in runner + Power lost due to mechanical resistance

270.756 = 208 + 51.9 + Power lost in runner + 4.3

Power lost in runner = 6.556 kW

3. A Pelton turbine having 1.6 m bucket diameter develops a power of 3600 KW at 400 rpm, under a net head of 275 m. If the overall efficiency is 88% and the coefficient of velocity is 0.97, let us determine: speed ratio, discharge, diameter of the nozzle and specific speed.

Solution:

Given data:

- $D = 1.6$ m
- $SP = 3600$ kW
- $N = 400$ rpm
- $H = 275$ m
- $\eta_o = 88\%$
- $C_v = 0.97$

To Find:

- Speed Ratio.
- Discharge.
- Diameter of the Nozzle and Specific Speed.

Formula to Be Used:

- $u = \frac{\pi DN}{60}$

- $$\eta_o = \frac{SP}{WP} = \frac{SP}{\frac{\rho g \times Q \times H}{1000}}$$

- $$Q = A \times V$$
  $$V_1 = C_v \sqrt{2gH}$$

- $$N_s = \frac{N\sqrt{Q}}{(H_m)^{3/4}}$$

Speed ratio:

$$u = \frac{\pi DN}{60}$$

$$u = \frac{\pi \times 1.6 \times 400}{60}$$

$$u = 33.51 \text{ m/s}$$

$$u = \phi\sqrt{2gH}$$

$$33.51 = \phi\sqrt{2 \times 9.81 \times 275}$$

$$\phi = 0.45$$

Discharge:

$$\eta_o = \frac{SP}{WP} = \frac{SP}{\frac{\rho g \times Q \times H}{1000}}$$

$$0.88 = \frac{3600}{\frac{100 \times 9.81 \times Q \times 275}{1000}}$$

$$Q = 1.516 \text{ m}^3/\text{s}$$

Diameter of Nozzle:

$$Q = A \times V$$

$$V_1 = C_v \sqrt{2gH} = 0.97 \times \sqrt{2 \times 9.81 \times 275}$$

$$V_1 = 71.25 \text{ m/s}$$

$$1.516 = \frac{\pi}{4} d^2 \times 71.25$$

$$d = 0.16 \text{m}$$

Specific Speed:

$$N_s = \frac{N\sqrt{Q}}{(H_m)^{3/4}}$$

$$= \frac{400 \times \sqrt{1.516}}{(275)^{3/4}}$$

$$N_S = \frac{492.50}{67.53}$$

$$N_S = 7.29 \text{rpm}$$

6.2.2 Francis Turbine

In Francis Turbine water flow is radial into the turbine and exits the turbine axially. Water pressure decreases as it passes through the turbine, imparting reaction on the turbine blades making the turbine rotate. Francis Turbine is the first hydraulic turbine with radial inflow. It was designed by American scientist James Francis.

Francis Turbine is a reaction turbine. Reaction turbines have some primary features which differentiate them from Impulse Turbines. The major part of pressure drop occurs in the turbine itself, unlike the impulse turbine where complete pressure drop takes place up to the entry point and the turbine passage is completely filled by the water flow during the operation.

Design of Francis Turbine

Francis Turbine has a circular plate fixed to the rotating shaft perpendicular to its surface and passing through its center. This circular plate has curved channels on it. The plate with channels is collectively called as runner. The runner is encircled by a ring of stationary channels called as guide vanes. Guide vanes are housed in a spiral casing called as volute.

The exit of the Francis turbine is at the center of the runner plate. There is a draft tube attached to the central exit of the runner. The design parameters such as radius of the runner, curvature of channel, angle of vanes and the size of the turbine as whole depend on the available head and type of application altogether.

Working of Francis Turbine

Francis Turbines are generally installed with their axis vertical. Water with high head (pressure) enters the turbine through the spiral casing surrounding the guide vanes. The water loses a part of its pressure in the volute (spiral casing) to maintain its speed. Then water passes through guide vanes where it is directed to strike the blades on the runner at optimum angles.

As the water flows through the runner its pressure and angular momentum reduces. This reduction imparts reaction on the runner and power is transferred to the turbine shaft, turbine is operating at the design conditions and the water leaves the runner in axial direction. Water exits the turbine through the draft tube which acts as a diffuser and reduces the exit velocity of the flow to recover maximum energy from the flowing water.

Power Generation using Francis Turbine

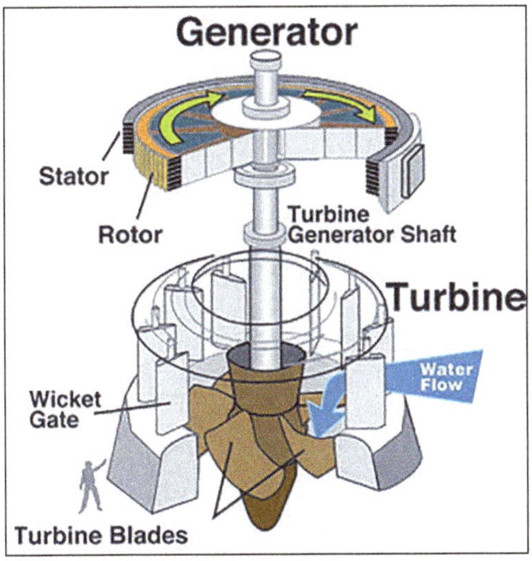

Francis turbine.

For power generation using Francis Turbine, the turbine is supplied with high pressure water which enters the turbine with radial inflow and leaves the turbine axially through the draft tube. The energy from water flow is transferred to the shaft of the turbine in the form of torque and rotation.

The turbine shaft is coupled with dynamos or alternators for power generation. For quality power generation speed of turbine should be maintained constant despite the changing loads. To maintain the runner speed constant even in reduced load condition the water flow rate is reduced by changing the guide vanes angle.

Problem

A reaction turbine works at 450 rpm under a heat of 120 m. Its diameter at inlet is 120 cm and the flow area is 0.4m2. The angles made by absolute and relative velocity at inlet are 20° and 60° respectively, with the tangential velocity. Let us determine the volume flow rate, the power developed and the hydraulic efficiency.

Solution:

Given:

- Speed of turbine: $N = 450$ r.p.m
- Head: $H = 120$ m
- Diameter at inlet: $d_1 = 120 = \text{cm} = 1.2$ m

To Find: The volume flow rate, the power developed and the hydraulic efficiency.

Formula to be used:

- $u_1 = \frac{\pi D_1 N}{60}$
- $Q = \pi \times D_1 \times B_1 \times V_{f_1}$
- $\rho \cdot Q\left[V_{w_1} u_1\right] \; (\because V_{W_2} = 2)$
- $\eta_h = \frac{V_{w_1} u_1}{gH} = \frac{35.79 \times 28.27}{9.81 \times 120}$

Flow Area $\pi d_1 \times B_1 = 0.4\,m^2$

Angle made by absolute velocity at inlet, $\alpha = 20°$

Angle made by the relative velocity at inlet, $G = 60°$

Whirl at outlet:

$$V_{w_2} = 0$$

Tangential velocity of the turbine at inlet:

$$u_1 = \frac{\pi D_1 N}{60} = \frac{\pi \times 1.2 \times 4.50}{60} = 28.27\,m/s$$

From inlet velocity triangle:

$$\tan\alpha = \frac{V_{f_1}}{V_{w_1}} \text{ or } \tan 20° = \frac{V_{f_1}}{V_{w_1}} \text{ or } \frac{V_{f_1}}{V_{w_1}} = \tan 20° = 0.364$$

$$\therefore V_{f_1} = 0.364\,V_{w_1} \qquad ...(1)$$

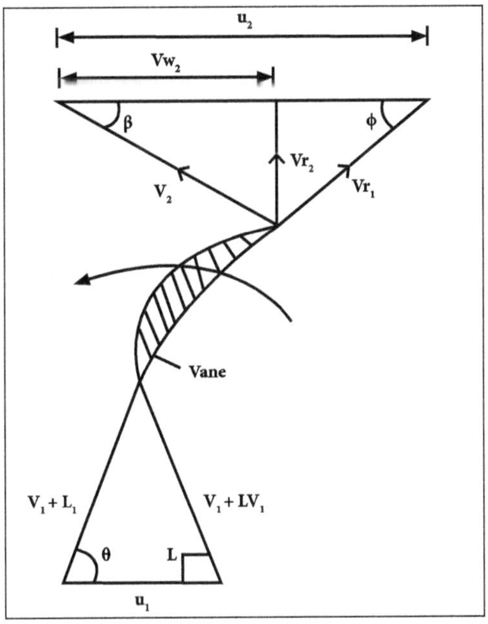

Also,

$$\tan\theta = \frac{V_{f_1}}{V_{w_1} - u_1} = \frac{0.364\,V_{w_1}}{V_{w_1} - 28.27} \quad (\because V_{f_1} = 0.364\,V_{w_1})$$

Or,

$$\frac{0.364\,V_{w_1}}{V_{w_1} - 28.27} = \tan\theta = \tan 60° = 1.732$$

$$\therefore 0.364\,V_{w_1} = 1.732(V_{w_1} - 28.27) = 1.732\,V_{w_1} - 48.96$$

Or,

$$(1.732 - 0.364)\,V_{w_1} = 48.96$$

$$\therefore V_{w_1} = \frac{48.96}{(1.732 - 0.364)} = 35.789 = 35.79 \text{ m/s}$$

From equation (1):

$$V_{f_1} = 0.364 \times V_{w_1} = 0.364 \times 55.79 = 13.027\,\text{m/s}$$

(i) Value flow rate is given by:

$$Q = \pi \times D_1 \times B_1 \times V_{f_1}$$

But,

$$\pi d_1 \times B_1 = 0.4\ \text{m}^2, Q = 0.4 \times 13.027 = 5.211\ \text{m}^3/\text{s}$$

(ii) Work done per second on the turbine is given by:

$$= \rho \cdot Q\left[V_{w_1} u_1\right] \left(\because V_{W_2} = 2\right)$$

$$= 1000 \cdot 5.211\ [35.79 \times 28.27] = 5272402\ \text{N m/s}$$

∴ Power developed in KW:

$$= \frac{\text{Work done per second}}{1000} = \frac{5272402}{1000} = 5272.402\ \text{KW}$$

(iii) The hydraulic efficiency is given by:

$$\eta_h = \frac{V_{w_1} u_1}{gH} = \frac{35.79 \times 28.27}{9.81 \times 120} = 0.8595 = 85.95\%$$

6.2.3 Kaplan Turbine

Working of Kaplan Turbine

The Kaplan turbine is a great development of early 20th century. It was invented by Prof. Viktor Kaplan of Austria. The Kaplan is of the propeller type, similar to an airplane propeller. The difference between the Propeller and Kaplan turbines is that the Propeller turbine has fixed runner blades while the Kaplan turbine has adjustable runner blades.

It is a pure axial flow turbine. The Kaplan's blades are adjustable for pitch and will handle a great variation of flow very efficiently. They are 90% or better in efficiency and are used in place some of the old Francis types in a good many of installations.

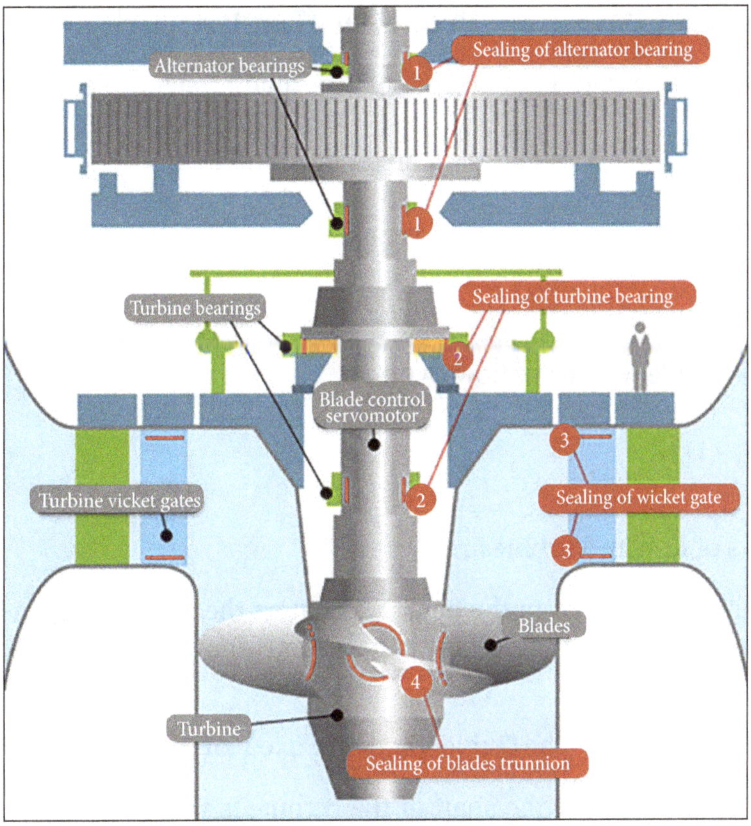

Kaplan turbine.

They are very expensive. In Kaplan turbine the runner's blades are movable. The application of Kaplan turbines are from a head of 2m to 40m.

Working

The water from the pen stock enters the scroll casing and then moves to the guide vanes. From the guide vanes, the water turns through 90° and flow axially through the runner. The discharge through the runner is obtained as:

$$Q=\frac{\pi}{4}\left(D_o^2-D_b^2\right)\times V_{f_1}$$

Where,

- D_o – Outer diameter of the runner.
- D_b – Diameter of hub.
- V_{f1} –Velocity of flow at inlet.

The inlet and outlet velocity triangles are drawn at the extreme edge of the runner vane corresponding to the points.

Some Important Points for Propeller (Kaplan Turbine)

1) The peripheral velocity @ inlet and outlet are equal.

$$\therefore u_1 = u_2 = \frac{\pi D_0 N}{60},\ D_0 \text{ -Outer diameter of runner.}$$

2) Velocity of flow at inlet and outlet are equal.

$$\therefore V_{f_1} = V_{f_2}$$

3) Area of flow @ inlet = Area of flow @ outlet.

$$= \frac{\pi}{4}\left(D_0^2 - D_b^2\right)$$

The main parts of Kaplan Turbine are:

1. Scroll Casing: The water from the penstocks enters the scroll casing and then moves to the guide vanes. From the guide vanes, the water turns through 90° and flows axially through the runner.

2. Guide Vane Mechanism: The Guide Vanes are fixed on the Hub.

3. Hub: For Kaplan Turbine, the shaft of the turbine is vertical. The lower end of the shaft is made larger and is called 'Hub' or 'Boss'. The vanes are fixed on the hub and hence Hub acts as runner for axial flow turbine.

4. Draft Tube: The pressure at the exit of the runner of Reaction Turbine is generally less than atmospheric pressure. The water at exit cannot be directly discharged to the tail race. The tube or pipe of gradually increasing area is used for discharging water from the exit of turbine to the tail race. This tube of increasing area is called Draft Tube. One end of the tube is connected to the outlet of runner while the other end is submerged below the level of water in the tail-race.

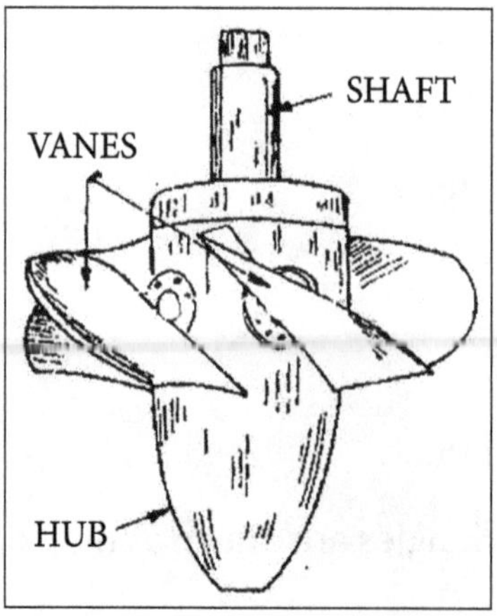

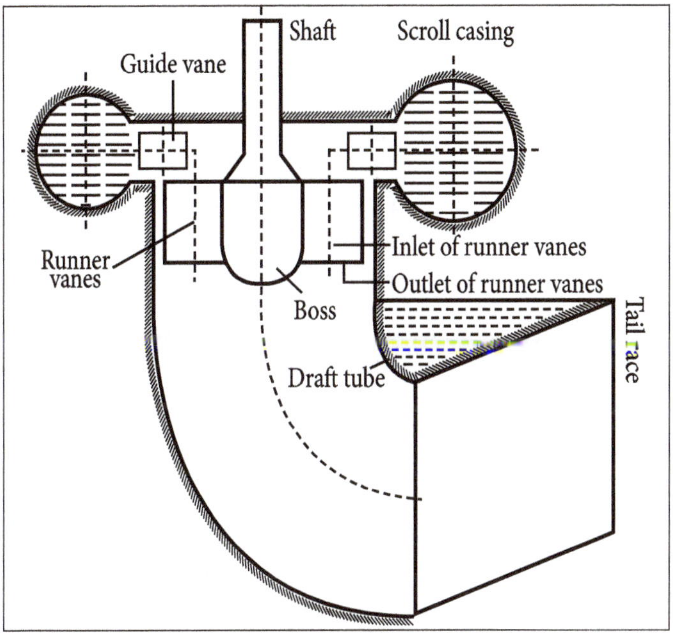

Draft Tube.

Advantages

- Very small no of blades are used nearly 3 to 8 blades.
- Less resistance has to be overcome.
- Runner vanes are adjusted in the Kaplan.
- Very low heads are required.

Disadvantages

- Speed of the turbine is 250 to 850.
- Position of the shaft is only in vertical direction.
- Large Flow rate must be required.
- High speed generator is required.

Problems

1. A Kaplan turbine runner is to be designed to develop 7357.5 kW shaft power. The net available head is 5.50 m. The speed ratio is 2.09, flow ratio is 0.68 and the overall efficiency is 60%. The diameter of the boss is 1/3rd of the diameter of the runner. Let us determine the diameter of the runner, its speed and its specific speed and solver it.

Solution:

Given:

- Shaft power, $p = 7357{\cdot}5$ kw
- Head, $H = 5{\cdot}5$ cm
- Speed ratio, $K_u = 2.09$
- Flow ratio, $K_f = 0.68$

To find: The diameter of the runner, its speed and its specific speed.

Formula to be used:

$$K_u = \frac{u_1}{\sqrt{2gH}} \Rightarrow u_1 = 2.09 \times \sqrt{2 \times 9.81 \times 5.50}$$

$$K_u = 21.71\,m/s$$

$$K_f = \frac{V_{f_1}}{\sqrt{2gH}} \Rightarrow V_{f_1} = 0.68\sqrt{2 \times 9.81 \times 5{\cdot}5}$$

$$V_{f_1} = 7.06\,m/s$$

$$\text{Overall efficiency}, e_o = \frac{\text{Shaft Power}(P)}{\text{Water Power}} = \frac{7357{\cdot}5}{WQH}$$

$$Q = \frac{7357{\cdot}5}{0.60 \times 9.81 \times 5.50}$$

$$Q = 227.27\,m^3/s$$

$$Q = A \times V_1$$

$$Q = \frac{\pi}{4}\left(D_o^2 - D_b^2\right) \times V_{f_1}$$

$$227.27 = \frac{\pi}{4}\left(D_o^2 - 1/9 D_o^2\right) \times 7.06$$

$$D_o^2 = 11.57$$

$$D_o = 3{\cdot}39\,m$$

$$u_1 = \frac{\pi D_o N}{60}$$

$$N = \frac{60 \times (u_1)}{\pi \times D_o} = \frac{60 \times 21.71}{\pi \times 3.39}$$

$$N = 122.31 \text{ rpm}$$

$$\text{Specific Speed}, N_s = \frac{N\sqrt{P}}{H^{5/4}}$$

$$= \frac{122.31 \times \sqrt{7357.5}}{(5.50)^{5/4}}$$

$$= \frac{10491.24}{8.423}$$

$$N_s = 1245.58$$

Diameter of boss $(D_b) = 1/3$ diameter of runner (Do)

Overall efficiency, $\eta_o = 60\%$

2. Let us calculate the diameter and speed of the runner of a Kaplan turbine developing 6000 kW under an effective head of 5 m. Overall efficiency of the turbine is 90%. The diameter of the boss is 0.4 times the external diameter of the runner. The turbine speed ratio is 2.0 and flow ratio 0.6.

Solution:

Given:

- Shaft power, $S.P. = 6000$ kW
- Head, $H = 5$ m
- Speed ratio = 2.0
- Flow ratio $= 0.6$
- Overall efficiency, $\eta_o = 90\% = 0.90$

Formula to be used:

Diameter of boss = 0.4 × external diameter

$$D_b = 0.4 \times D_o$$

Speed ratio:

$$u_1 \sqrt{2gh} = 2.0$$

$$u_1 = 2 \times \sqrt{2gh} = 2.0 \times \sqrt{2 \times 9.81 \times 5}$$

$$u_1 = 19.8 \, m sec$$

Flow ratio:

$$\frac{V_{f_1}}{\sqrt{2gH}} = 0.6$$

$$V_{f_1} = 0.6\sqrt{2gH}$$

$$= 0.6\sqrt{2 \times 9.81 \times 5}$$

$$= 5.94 \, m / sec$$

$$\eta_o = \frac{S.P}{W.P}$$

$$0.90 = \frac{6000}{\frac{\rho \times g \times Q \times H}{1000}}$$

$$0.90 = \frac{6000}{\frac{1000 \times 9.81 \times Q \times 5}{1000}}$$

$$Q = \frac{6000 \times 1000}{1000 \times 9.81 \times 5 \times 0.90}$$

$$Q = 136.91 \, m^3 / s$$

$$Q = \pi / 4\left(D_o^2 - D_b^2\right) \times V_{f_1}$$

$$135.91 = \pi / 4\left[D_o^2 - \left(0.4D_o\right)^2\right] \times 5.94$$

$$135.91 = 3.92 D_o^2$$

$$D_o^2 = \frac{135.91}{3.92}$$

$$D_o = \sqrt{\frac{135.91}{3.92}}$$

$$D_o = 5.88\,m$$

$$D_b = 0.4 \times D_o = 0.4 \times 5.88$$

$$D_b = 2.352\,m$$

Speed of the turbine is given by $u1 = \frac{\pi D_o N}{60}$

$$19.8 = \frac{11 \times 5.88 \times N}{60}$$

$$N = 64.31 \text{ rpm}$$

6.3 Hydraulic Design: Draft Tube, Theory, Functions and Efficiency

Draft Tube

The draft tube connects the runner exit to the tail race where the water from the turbine is being discharged finally. The primary function of the draft tube is to reduce the velocity of the discharged water in order to minimize the kinetic energy loss at the outlet. This allows the turbine to be set above the tail water without any appreciable drop of available head. A clear understanding of the draft tube function is important for the purpose of its design.

Let us obtain an expression for the work done per second by water on the runner of a Pelton wheel and expression for maximum efficiency of the Pelton wheel giving the relationship between the jet speed and bucket speed.

Let,

H = Net head acting on the Pelton wheel

$$H = H_g - h_f$$

Where, H_g = Gross head and $h_f = \frac{4fLV^2}{D^* X_{2g}}$.

Where,

- D^* = Dia. of Pen stock.
- N = Speed of the wheel in r.p.m.

- D = Diameter of the wheel.
- d = Diameter of the jet.

Then,

$$V_1 = \text{Velocity of jet at inlet} = \sqrt{2gH}$$

$$u = u_1 = u_2 = \frac{\pi DN}{60}$$

The velocity triangle at inlet will be a straight line where:

$$V_{r_1} = V_1 - u_1 = V_1 - u$$

$$V_{W_1} = V_1, \alpha = 0 \text{ and } \theta = 0$$

From the velocity triangle at outlet, we have:

$$V_{r_1} = V_{r_1} \text{ and } V_{W_2} = V_{r_2} \cos\phi - u_2$$

The force exerted by the jet of water in the direction of motion is given by equation as:

$$F_X = \rho a V_1 \left[V_{W_1} + V_{W_2} \right]$$

As the angle β is an acute angle, +ve sign should be taken. Also this is the case of series of varies; the mass of water striking is $\rho\alpha V_1$ and not $\rho a v_{r1}$. In equation, 'a' is the area of the jet which is given as:

$$a = \text{Area of Jet} = \frac{\pi}{4} d^2$$

Now work done by the jet on the winner per second:

$$= F_X \times u = \rho a V_1 \left[V_{W_1} + V_{W_2} \right] \times u \text{ Nm/s}$$

Power gives to the runner by the jet:

$$= \frac{\rho a V_1 \left[V_{W_1} + V_{W_2} \right] \times u}{1000} \text{KW}$$

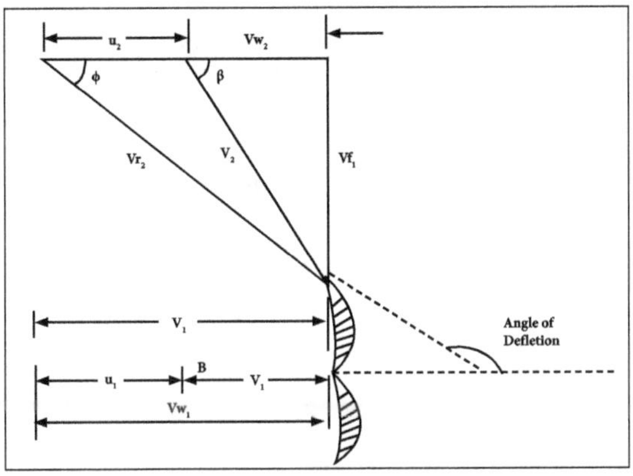

Work done/s per unit weight of water Striking/s:

$$= \frac{\rho a V_1 \left[V_{w_1} + V_{w_2} \right] \times u}{\text{Weight of Water Strikings}}$$

$$= \frac{\rho a V_1 \left[V_{w_1} + V_{w_2} \right] \times u}{\rho a V_1 \times 8} = \frac{1}{8} \left[V_{w_1} + V_{w_2} \right] \times u$$

The energy supplied to the jet at inlet is in the form of kinetic energy and is equal to $\frac{1}{2} mV^2$.

$$\therefore \text{ K. E of jet per second } = \frac{1}{2} (\rho a V_1) \times V_1^2$$

$$= \frac{\rho a V_1 \left[V_{w_1} + V_{w_2} \right] \times u}{\frac{1}{2} (P a V_1) \times V_1^2} = \frac{2 \left[V_{w_1} + V_{w_2} \right] \times u}{V_1^2}$$

$$\therefore \text{ Hydraulic efficiency, } \eta_h = \frac{\text{Work done per second}}{\text{K.E. of jet per second}}$$

Now,

$$V_{w_1} = V_1, V_{r_1} = V_1 - u_1 = (V_1 - u)$$

$$\therefore Vr_2 = (V_1 - u)$$

And,

$$V_{w_2} = V_{r_2} \cos\phi - u_2 = V_{r_2} \cos\phi - u = (V_1 - u)\cos\phi - 1$$

Substituting the values of V_{w1} and V_{W2} in equation:

$$\eta_h = \frac{2\left[V_1 + (V_1 - u)\cos\phi - u \right] \times u}{v_1^2}$$

$$=\frac{2\left[V_1-u+(V_1-u)\cos\phi\right]\times u}{V_1^2}=\frac{2(V_1-u)[1+\cos\phi]\times u}{V_1^2}$$

The efficiency will be maximum for a given value of V1 when:

$$\frac{d}{du}(\eta_h)=0 \text{ or } \frac{d}{du}\left[\frac{2u(v_1-u)(1+\cos\phi)}{v_1^2}\right]=0$$

Or,

$$\frac{(1+\cos\phi)}{v_1^2}\frac{d}{du}\left(2uV_1-2u^2\right)=0 \text{ or } \frac{d}{du}\left[2uV_1-2u^2\right]=0$$

$$\left(\because\frac{1+\cos\phi}{v_1^2}\neq 0\right)$$

Or,

$$2V_1-4u=0 \text{ or } u=\frac{V_1}{2}$$

Equation states that hydraulic efficiency of a Pelton wheel will be maximum when the velocity of the wheel is half the velocity of the jet of water at inlet. The expression for maximum efficiency will be obtained by substituting the value of $=\frac{V_1}{2}$ in equation:

$$\therefore \text{Max.}\eta_h=\frac{2\left(V_1-\frac{V_2}{2}\right)(1+\cos\phi)\times\frac{V_1}{2}}{V_1^2}$$

$$=\frac{2\times\frac{V_1}{2}(1+\cos\phi)\frac{V_1}{2}}{V_1^2}=\frac{(1+\cos\phi)}{2}$$

6.4 Geometric Similarity: Unit and Specific Quantities

6.4.1 Unit Quantities

In order to predict the behavior of a turbine working under varying conditions of head, speed and power, recourse has been made to the concept of unit.

The unit quantities give the speed, discharge and power for a particular turbine under a head of 1m assuming the same efficiency. The following are the three important unit quantities:

- Unit speed.
- Unit power.
- Unit discharge.

Unit speed (N_u)

The speed of the turbine, working under unit head (say 1m) is known as unit speed of the turbine.

The tangential velocity is given by:

$$u=\frac{\pi DN}{60} \text{ or } N=\frac{60u}{\pi D}$$

$$\text{If } H=1; \text{ then } N=N_u\sqrt{H}$$

Where, H = head of water, under which the turbine is working, N= speed of turbine under a head, H, u= tangential velocity, Nu= speed of the turbine under a unit head.

Unit Power (P_u)

The power developed by a turbine, working under a unit head (say 1m) is known as unit power of the turbine.

Power developed by a given as:

$$P=\gamma QH \text{ and } V=\sqrt{2gH}$$

$$P=\gamma\left(a\sqrt{2gH}\right)H$$

$$P=K_2H^{3/2}$$

$$\text{If } H=1; \text{then}, P=P_u$$

$$P_u=K_2 1^{3/2}=K_2$$

$$P=P_uH^{3/2}$$

$$\text{Thus}, P_u=\frac{P}{H^{3/2}}$$

Unit Discharge (Q_u)

The discharge of the turbine working under a unit head (say 1m) is known as unit discharge.

$$Q = a\sqrt{2gH} = K_3\sqrt{H}$$

$$\text{If } H=1;\text{then}, Q=Q_u$$

$$Q_u = K_3\sqrt{1} = K_3$$

Or,

$$Q = Q_u\sqrt{H}$$

Thus,

$$Q_u = \frac{Q}{\sqrt{H}}$$

If a turbine is working under different heads, the behavior of the turbine can be easily known from the unit quantities.

$$N_u = \frac{N_1}{\sqrt{H_1}} = \frac{N_2}{\sqrt{H_2}}$$

$$P_u = \frac{P_1}{H_1^{\frac{3}{2}}} = \frac{P_2}{H_2^{\frac{3}{2}}}$$

$$Q_u = \frac{Q_1}{\sqrt{H_1}} = \frac{Q_2}{\sqrt{H_2}}$$

6.4.2 Characteristic Curves

The following are the important characteristic curves of a turbine:

Main Characteristic Curve

- Operating characteristic curve.
- Constant efficiency curve.

Main characteristic curve are obtained by maintaining a constant head and a constant gate opening (G.O) on the turbine. The speed of the turbine is varied by changing load on the turbine. For each value of the speed, the corresponding value of power (P) and discharge (Q) are obtained. Then the overall efficiency (η_o) for each value of the speed is calculated.

From these readings the value of unit speed (N_U) unit power (P_U) and unit discharge (Q_U) are determined. Taking Nu as abscissa, the values of Q_u, P_u, P and η_o are plotted. By changing the gate opening, the values of Qu, Pu and ηo and Nu are determined are taking Nu as abscissa the value of Q_u, P_u and η_o are plotted.

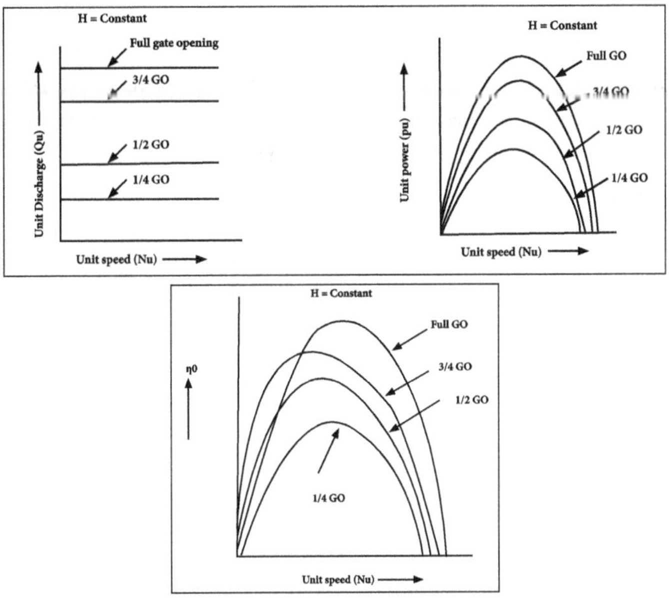

Main characteristic curve.

Operating Characteristic Curves

The operating characteristic curves are plotted when the speed on turbine is constant. In case of turbines, the head is generally constant. There are three independent parameters namely N, H and Q. For operating characteristics N and H are constant and hence the variation of power and efficiency with respect to discharge Q are plotted. The power and efficiency curves will be slightly away from origin on x-axis, as to overcome initial friction certain amount of discharge will be required.

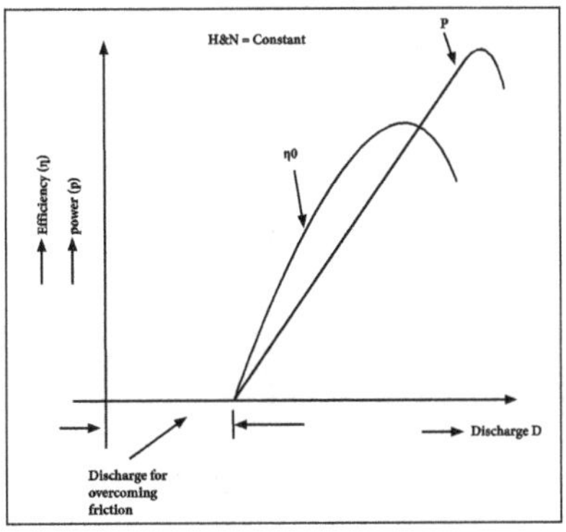

Operating characteristic curve.

Constant Efficiency Curves

These current are obtained from speed Vs efficiency and speed Vs discharge curves for different gate opening. For a given efficiency from Nu Vs ηo curves, there are two speeds. From Nu Vs Qu versus, corresponding to two values of speeds there are two values of discharge. Hence for a given efficiency there are two values of discharge for a particular gate opening.

This means for a given efficiency there are two values of speeds and two values of discharge for a given gate opening. If the efficiency is maximum there is only one value. These two values of speed and two values of discharge corresponding to a particular gate opening are plotted. The procedure is repeated for different gate opening and the curve Q Vs N are plotted.

The points having the same efficiency are joined. The curves having same efficiency are called is co-efficiency curves. These curves are helpful for determining the zero of constant efficiency and for predicating the performance of turbine at various efficiencies.

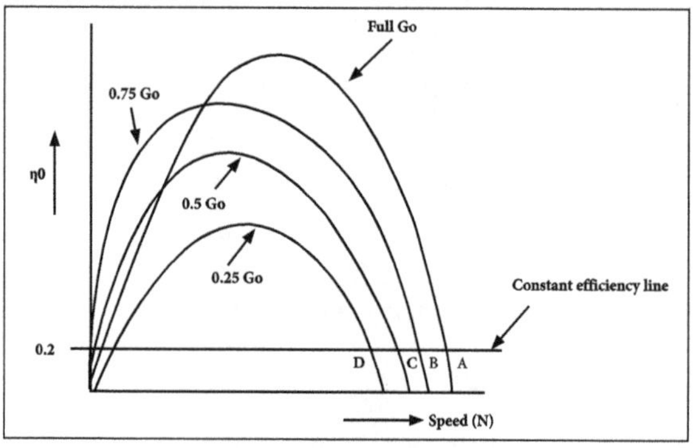

Constant Efficiency Curve.

6.5 Governing of Turbines

Governor is the heart of a turbine which controls it during steady state and transient conditions. The governor does this function by controlling the steam flow through the turbine by adjusting the control value. The governor is a control component with advanced protections for the turbine which ensure safe operation of the turbine.

Governing system of the turbine does the following functions:

- Controls the turbine speed during start-up or in no load condition to permit the unit to be synchronized with the grid.
- Controls the turbine load when running in parallel with the grid/generating sets.
- Provides all protective functions to ensure the safe operation of the unit.

Methods of Governing

Depending on the final control component of governing system. These are divided as nozzle governing and throttle governing.

Nozzle Governing

This type of governing is suitable for simple impulse turbine and most efficiently in part load conditions. In this method the steam flow is regulated by opening or closing some set of nozzles of the turbine. These nozzles are actuated by independent control mechanisms. Depending upon the load demand some number of nozzles will be put in service and some put out of service.

This feature enables the use of nozzle governing for part load operation efficiently. For example, if the turbine consists of 10 nozzles for full power (100 %). If the load demand has reduced to 60 % then out of 10 nozzles, 4 nozzles can be closed to provide 60% load without delay. This gives the advantage on throttle governing where the power reduction will happen gradually.

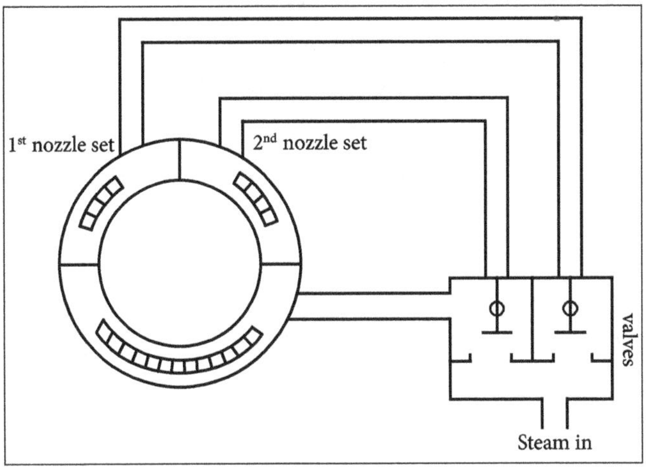

Nozzle governing mechanism of steam turbine.

In nozzle governing method, there will not be any importance to the steam inlet pressure and it is done by the boiler and bypass pressure control system. So the use of nozzle governing is limited to small turbines.

Throttle Governing

In this case of governing mechanism, the mass flow rate is controlled by throttling the steam by means of control valve. The control valve position will be adjusted to allow the required steam for turbine operation. Here the controlling parameter can be turbine speed and load. With latest technologies steam generator pressure control is also incorporated in governing control system. This is more important in nuclear power plants where constant steam pressure is essential for efficient operation.

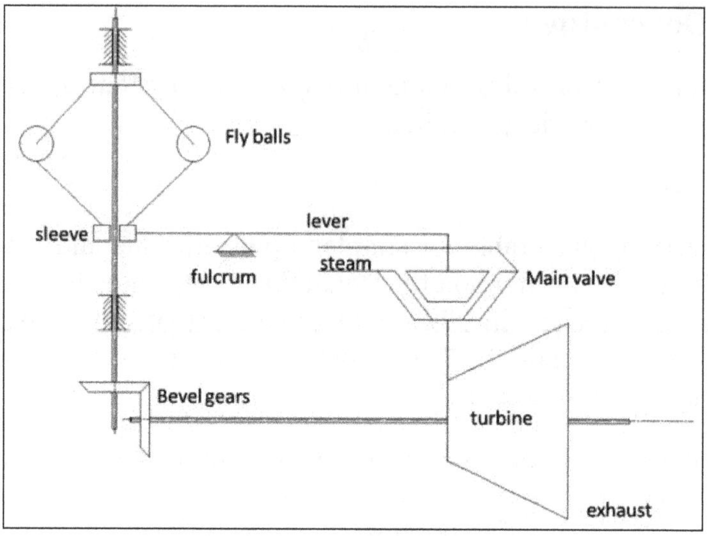

Throttling sleeve governor of steam turbine.

For example, if the plant is operating at full power and it needs to be operated at 60 % according to the load demand. Then the controller checks the demand with the existing load and gives a adjustment signal to control valve. The control valve will slowly throttle to reduce the steam flow for acquiring the load demand.

Due to throttling of control valves there will be high energy loss across the control valves. Hence the design of valve should be adequate to withstand the pressure of the steam.

6.6 Selection of Type of Turbine, Cavitation, Surge Tank and Water Hammer

6.6.1 Selection of Type of Turbine

The following points should be considered while selecting right type of turbines:

1. Specific speed: High specific aped is essential where the head is low and output is large. Otherwise the rotational speed will be low which means cost of turbo-generator and powerhouse will be high. On the other hand there is no need of choosing a high value of specific speed for high installations, because with low specific speed high rotational speed can be attained.

2. Rotational speed: Rotational speed depends upon specific speed. The rotational speed of an electrical generator depends on the frequency and number of pair of poles. The value of specific speed adopted should be so that it will give the synchronous speed of the generator.

3. Cavitation: The installation of reaction turbines is affected by cavitation. The critical values of cavitation indices must be obtained to see that the turbine works in safe zone. Such values of cavitation indices also affect the design of turbine, especially of Kaplan, propeller and bulb types.

4. Deposition of turbine shaft: The vertical shaft arrangement is better for large-sized reaction turbines. Therefore it is universally adopted whereas in case of large size impulse turbines, horizontal shaft arrangement is preferable.

5. Efficiency: The efficiency selected should be such that it gives the highest overall efficiency of various conditions.

6. Part load operation: In general the efficiency at part loads and overloads is less than that with rated (design) parameters. For the sake of economy the turbine should always run with maximum possible efficiency to get more revenue. When the turbine has to run at part or overload conditions Deriaz turbine is employed. Similarly, for low heads, Kaplan turbine will be useful for such purposes in place of propeller turbine.

7. Available head and its fluctuation:

- Very high (350 m and above): A greater heads than 350m, Pelton Turbine is generally employed and practically there is no any choice except in very special cases.
- High heads (150 m to 350 m): In this range either Pelton or Francis turbine may be employed. For higher specific needs Francis turbine is more compact and economical than the Pelton turbine.
- Medium heads (60 m to 150 m): A Francis turbine is usually employed in this range. High or low specific speed would be used depending on the selection of the speed.
- Low heads (below 60 m): The between 30m to 60m both Kaplan and Francis turbines may be used. Francis is more expensive but yields higher efficiency at part loads and over loads. Propeller turbines are however, commonly used for heads up to 15m. They are adopted when there is practically no load variation.

8. Water quality: The Quality of water is more crucial for the reactive turbine the in reaction turbines. Reactive turbine may undergo for rapid wear in high head reactive turbines.

6.6.2 Cavitation

If the pressure in the liquid is less than its vapor pressure, the liquid will boil and bubbles of vapor will form. As the fluid flows into region of higher pressure the bubbles of vapor will suddenly condense or collapse. This produces a high dynamic pressure upon the adjacent walls and since the action is continuous there will be damage. Turbine runners and pump impellers are often severely damaged by such action. The process is called cavitation and the damage is called cavitations damage. In order to avoid cavitations, the absolute pressure at all points should be above the vapor pressure

6.6.3 Surge Tank

Surge tank (or surge chamber) is a device which is used to absorb the excess pressure rise in case of a sudden valve closure. The surge tank is located between the horizontal or slightly inclined conduit and steeply sloping penstock and is designed as a chamber excavated in the mountain.

It also acts as a small storage from which water may be supplied in case of a sudden valve opening of the turbine. In case of a sudden opening of turbine valve, there is a chance for penstock to collapse due to a negative pressure generation, if there is no surge tank.

Surge Tank Function

When the valve in a hydroelectric power plant is suddenly closed, the water in the penstock stops at once. The water in the pipeline with large inertia retards slowly. The difference in flows between pipeline and penstock causes a rise in the water level in the surge tank. The water level rises above the static level of the reservoir water, producing a counter-pressure. As a result water in the pipeline flows towards the reservoir and the level of water in the surge tank drops. In the absence of damping, oscillation would continue indefinitely with the same amplitude.

The flow into the surge tank and water level in the tank at any time during the oscillation depends on the dimension of the pipeline and tank and on the type of valve movement.

The main functions of a surge tank are:

- It reduces the amplitude of pressure fluctuations by reflecting the incoming pressure waves.
- It improves the regulation characteristic of a hydraulic turbine.

6.6.4 Water Hammer

Water hammer is caused by a shock or pressure wave that travels through the pipes which are generated by a sudden stop in the velocity of the water or a change in the direction of flow. If the pipe is suddenly closed at the outlet, the mass of water before the closure is still moving forward with some velocity, building up a high pressure and shock waves. This result in a loud noise called as water hammer.

Method to Control Water Hammer

The most effective means of controlling water hammer is the use of measured, compressible cushion of air which is permanently separated from the water system. Water hammer arresters employ a pressurized cushion of air and a two O-ring piston which can permanently separates this air cushion from the water system. When the valve closes and the water flow is suddenly stopped, the pressure spike pushes the piston up the arrester chamber against the pressurized cushion of air. The air cushion in the arrester instantly reacts by absorbing the pressure spike that causes water hammer.

6.7 Hydraulic Systems

6.7.1 Hydraulic Ram

Its operation depends on the phenomenon called water hammer. More than 50% of the energy of the driving flow can be transferred to the delivery flow. Initially the impulse valve will be open under gravity (or by a light spring) and water will flow down the drive pipe (through a strainer) from the water source. As the flow accelerates, the hydraulic pressure and the static pressure increases until the resulting forces overcome the weight of the impulse valve and start to close it.

As soon as the valve aperture decreases, the water pressure builds up rapidly and slams the impulse valve shut. The moving column of water in the drive pipe is no longer able to exit via the impulse valve so its velocity must suddenly decrease. This continues to cause a considerable rise of pressure which forces to open the delivery valve to the air chamber.

Once the pressure exceeds the static delivery head, water will be forced up the delivery pipe. Air trapped in the air chamber will be compressed simultaneously to a pressure exceeding the delivery pressure. Eventually the column of water in the drive pipe comes to a halt and the static pressure in the casing then falls to near the supply head pressure. The delivery valve will then close, when the pressure in the air chamber exceeds that in the casing.

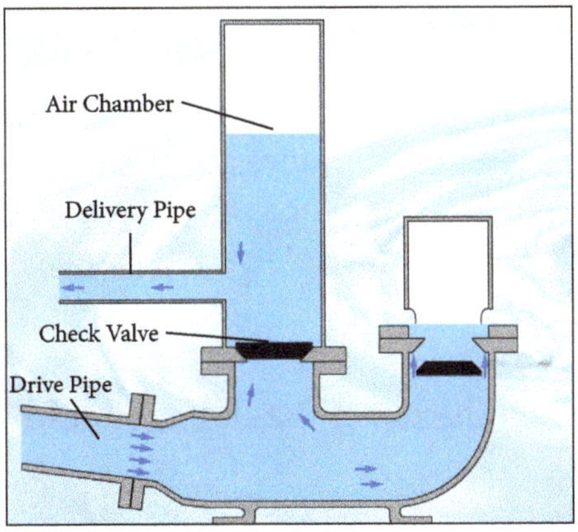

Hydraulic ram.

Water will continue to be delivered after the delivery valve has closed until the compressed air in the air chamber has expanded to a pressure equal to the delivery head. A check valve is included in the delivery pipe to prevent return flow. When the delivery valve closes, the reduced pressure will allow the impulse valve to drop under its own weight.

6.7.2 Hydraulic Lift

Hydraulic lift is classified are:

Direct Acting Hydraulic Lift:

- Fixed cylinder which is fixed with the wall of the floor, where the sliding ram reciprocate when we apply the pressure.
- Cage which fitted on the top of the sliding ram where the load is placed (i.e. lifted load).
- Sliding ram which is fitted in the fixed cylinder reciprocates (upward or downward direction) when we apply the pressure (i.e. reaches the floor wise.)

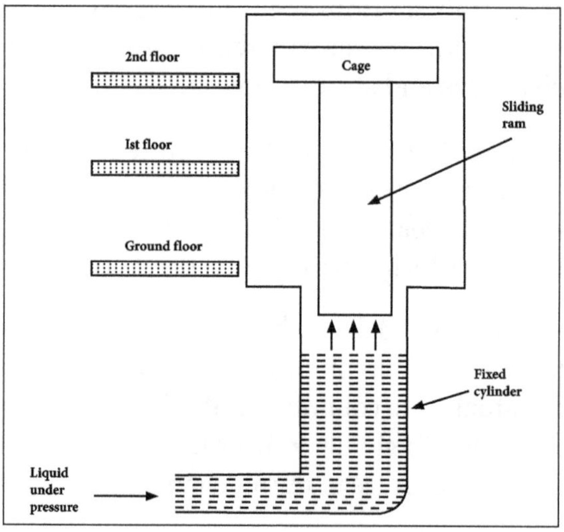

Direct Acting Hydraulic Lift.

When fluid under pressure is forced into the cylinder, the ram gets a push upward. The platform carries loads or passengers and moves between the guides. At required height, it can be made to stay in level with each floor so that the good or passengers can be transferred. In direct acting hydraulic lift, stroke of the ram is equal to the lift of the cage.

Suspended Hydraulic Lift:

- Cage which is fitted on the top of the sliding ram where the load is placed (i.e. lifted load).
- Wire rope which connects the cage to pulley.

- Sliding ram which is fitted in the fixed cylinder which is reciprocate (upward or downward direction) when we applied the pressure (i.e. reaches the floor wise).
- Pulleys which are connected to the sliding ram and fixed cylinder, where one pulley is fixed and other pulley is movable.
- Hydraulic jigger which consists of a moving ram which slides inside a fixed hydraulic cylinder.
- Fixed cylinder which is fixed with the wall of the floor, where the sliding ram reciprocate when we apply the pressure.

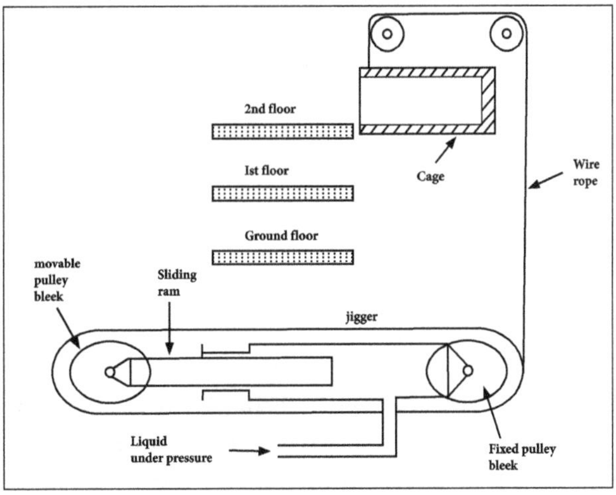

Suspended Hydraulic Lift.

When fluid under pressure is forced into the cylinder, the ram gets reciprocate to the movable pulleys. With the help of arrangement of hydraulic jigger, pulley can rotate, with the help of wire rope the cage is maintain their pressure force with their floor. At required height, it can be made to stay in level with each floor so that the good or passengers can be transferred.

Working period of the lift is ratio of the height of lift to the velocity of lift. Idle period of lift is the difference of the total time for one operation and the working period of the lift.

6.7.3 Hydraulic Coupling

It is a device to transmit the rotation between shafts by means of the acceleration and deceleration of a hydraulic fluid. A hydraulic coupling consists of an impeller (driving shaft) on the input and a runner (driven shaft) on the output. They contain the fluid.

The impeller acts as a pump and the runner act as a turbine. Basically, the impeller accelerates the fluid from near its axis where the tangential component of absolute velocity is low to near its periphery where the tangential component of absolute velocity is high.

This increase in velocity represents an increase in kinetic energy. The fluid mass emerges at high velocity from the impeller, impinges on the runner blades, gives up its energy and leaves the runner at low velocity.

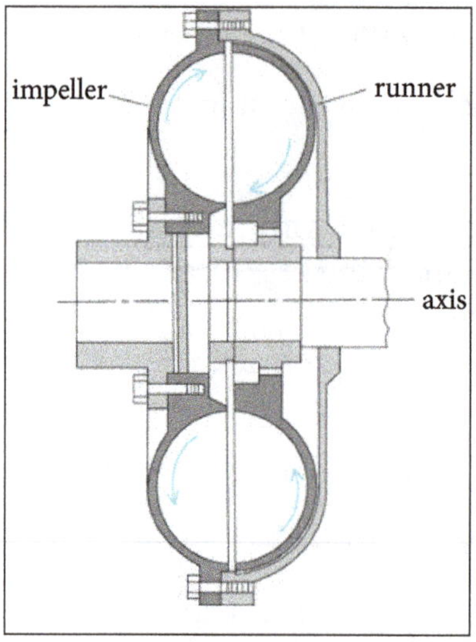

Hydraulic coupling.

Hydraulic couplings transfer rotational force from a transmitting axis to a receiving axis. The coupling consists of two toroid's (doughnut-shaped objects) in a sealed container of hydraulic fluid. One toroid is attached to the driving shaft and spins with the rotational force. The spinning toroid moves the hydraulic fluid around the receiving toroid. The movement of the fluid turns the receiving toroid and thus turns the connected shaft.

Even though fluid couplings use hydraulic fluid within their construction they lose a portion of force to friction and results in the creation of heat. No fluid coupling can run at 100 percent efficiency. Excessive heat production from poorly maintained couplings can cause damage to the coupling and surrounding systems.

6.8 Fluidics: Amplifiers, Sensors and Oscillators

Fluidics

Fluidics is defined as the technology which deals with the use of flowing liquid or gas in various devices to perform functions that are usually performed by an electric current in electronic devices.

Fluidic Amplifiers

Fluidic amplifiers are used to apply a gain in pressure, flow or power to a fluidic circuit. These amplifiers are analogous to voltage, current and power amplifiers used in electronic circuits. It considers the jet-deflection amplifier shown in the figure. When the control input pressures pu1 and pu2 are equal, the supply stream passes through the amplifier with a symmetric flow.

Then the output pressures py1 and py2 are equal. When $p_{u1} \neq p_{u2}$ the fluid stream is deflected to one side due to the nonzero differential pressure $\Delta p_u = p_{u1} - p_{u2}$. As a result a nonzero differential pressure $\Delta p_y = p_{y2} - p_{y1}$ is created at the output. The pressure gain of the amplifier is given by:

$$Kp = \frac{\Delta p_y}{\Delta p_u}$$

The gain K_p will be constant in a small operating range.

Fluidic Sensors

A fluidic displacement sensor [figure (a)] can be developed by using a mechanical vane to split a incoming flow stream into two output streams. When the vane is centrally located, the displacement is zero. Under these conditions, the pressure is same at both output streams with the differential pressure $p_2 - p_1$ remaining at zero. When the vane is not symmetrically located, the output pressures will be unequal. The differential pressure $p_2 - p_1$ provides both magnitude and direction of the displacement θ.

A fluidic angular speed sensor is shown in the figure (b). The nozzle of the input stream is rotated at angular speed ω. When $\omega = 0$, the stream travels straight in the axial direction of the input stream. When $\omega \neq 0$, the fluid particles emitting from the nozzle have a transverse speed as well as an axial speed due to jet flow. Hence the fluid particles are deflected from the original path. This deflection results in pressure change at the output. Hence the output pressure change can be used as a measure of the angular speed.

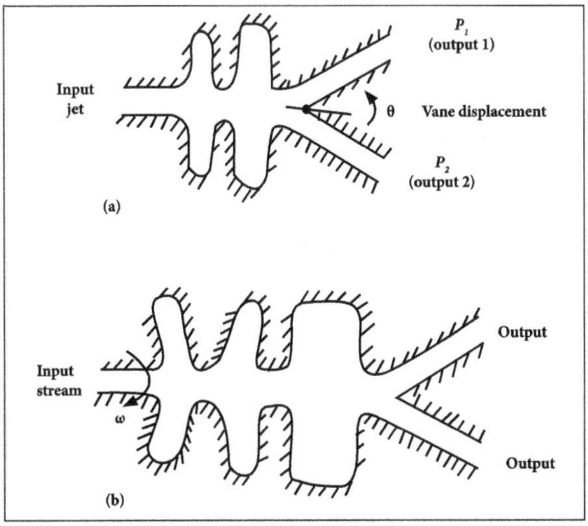

(a) Angular displacement sensor (b) Laminar angular speed sensor.

There are many ways to sense speed using a fluidic device. One type of fluidic angular speed sensor uses the vortex flow principle. In this sensor, the angular speed of the input device is imparted on the fluid entering a vortex chamber at the periphery.

In this manner a tangential speed is applied to the fluid particles, which moves radially from the periphery towards the vortex center. The resulting tangential speed becomes larger as the particles approach the center of the chamber.

A pressure drop is experienced at the output. The higher the angular speed imparted to the incoming fluid, the larger the pressure drop at the output. Hence the output pressure drop can be used as a measure of angular speed.

Fluidic Oscillators

Fluidic oscillators are small-scale devices that emit a spatially oscillating jet at high frequency and velocity up to 20 kHz. The fluidic oscillator ability is to impart an oscillation movement to a fluid flow at extremely high frequencies whereby the mixing process is improved. Fluidic oscillators can be used in a wide range of applications where the mixing process plays an vital role.

Function of a Fluidic Oscillator

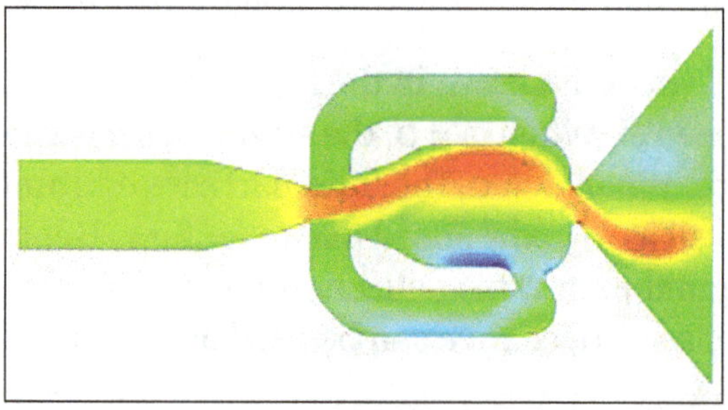

Fluidic Oscillator.

The supplied fluid is attached to either walls of the oscillator chamber when passing through the actuator. A small portion is diverted through the feedback path and it impinges the main jet at the inlet. This causes the main jet to switch to the opposite side of the chamber, triggering an oscillating flow. For small dimensions of the device, the frequency results in high range of 1000 Hz.

Advantages of Fluidics

- Reliable.
- Simple in construction.
- Since they contain no moving parts, not much maintenance problems are encountered.

- Mode of energy feeding to a fluidic system is very simple.
- Have good response and performance characteristics.
- Offer exceptional physical and thermal stability.
- Noise free.
- Smaller in size.

Disadvantages of Fluidics

- Unsuitable for incompressible fluids.
- Not suitable for intermittent operation control systems.
- Inefficient in operation.
- Slow speeds and low power outputs.
- Limited development of the field.
- Complex systems arc impracticable.

Applications of Fluidics

- To measure flow rates.
- To check weights.
- To operate various types of machinery, etc.
- To provide on-off controls.
- If used with a chemically inert gas such as nitrogen or helium, they can be used in the manufacture of explosives where fire or electrical hazards are present.

Permissions

All chapters in this book are published with permission under the Creative Commons Attribution Share Alike License or equivalent. Every chapter published in this book has been scrutinized by our experts. Their significance has been extensively debated. The topics covered herein carry significant information for a comprehensive understanding. They may even be implemented as practical applications or may be referred to as a beginning point for further studies.

We would like to thank the editorial team for lending their expertise to make the book truly unique. They have played a crucial role in the development of this book. Without their invaluable contributions this book wouldn't have been possible. They have made vital efforts to compile up to date information on the varied aspects of this subject to make this book a valuable addition to the collection of many professionals and students.

This book was conceptualized with the vision of imparting up-to-date and integrated information in this field. To ensure the same, a matchless editorial board was set up. Every individual on the board went through rigorous rounds of assessment to prove their worth. After which they invested a large part of their time researching and compiling the most relevant data for our readers.

The editorial board has been involved in producing this book since its inception. They have spent rigorous hours researching and exploring the diverse topics which have resulted in the successful publishing of this book. They have passed on their knowledge of decades through this book. To expedite this challenging task, the publisher supported the team at every step. A small team of assistant editors was also appointed to further simplify the editing procedure and attain best results for the readers.

Apart from the editorial board, the designing team has also invested a significant amount of their time in understanding the subject and creating the most relevant covers. They scrutinized every image to scout for the most suitable representation of the subject and create an appropriate cover for the book.

The publishing team has been an ardent support to the editorial, designing and production team. Their endless efforts to recruit the best for this project, has resulted in the accomplishment of this book. They are a veteran in the field of academics and their pool of knowledge is as vast as their experience in printing. Their expertise and guidance has proved useful at every step. Their uncompromising quality standards have made this book an exceptional effort. Their encouragement from time to time has been an inspiration for everyone.

The publisher and the editorial board hope that this book will prove to be a valuable piece of knowledge for students, practitioners and scholars across the globe.

Index

www.ingramcontent.com/pod-product-compliance
Lightning Source LLC
Chambersburg PA
CBHW062006190326
41458CB00009B/2986
9781647266479